Advanced Sorption Process Applications

Edited by Serpil Edebali

Published in London, United Kingdom

IntechOpen

Supporting open minds since 2005

Advanced Sorption Process Applications
http://dx.doi.org/10.5772/intechopen.75857
Edited by Serpil Edebali

Contributors
Mazen Nazal, Ifeoma Mary Ugwu, Onyedika Igbokwe, Kassio Mendes, Ananias Francisco Dias Júnior, Vanessa Takeshita, Ana Paula Justiniano Régo, Valdemar Tornisielo, Noureddine Abidi, Rohan Dassanayake, Sanjit Acharya, Márcio Antônio Fiori, Gustavo Colpani, Micheli Zanetti, Rubieli Carla Frezza Zeferino, Luciano Luiz Silva, Josiane Maria Muneron Mello, Augustine Muwamba, Kelly T. Morgan, Peter Nkedi-Kizza, Naoki Kano, Shuang Zhang, Serdar Aydin, Kurtulus Ozcan, Serkan Emik, Atakan Öngen, George William Kajjumba, Lizethly Caceres Jensen, Jorge Rodríguez Becerra, Mauricio Escudey, Marcelo Alves, Johannes Lützenkirchen, Yujun Wang, Ahmed Abdelmonem

First published in London, United Kingdom, 2019 by IntechOpen
IntechOpen is the global imprint of INTECHOPEN LIMITED, registered in England and Wales, registration number: 11086078, The Shard, 25th floor, 32 London Bridge Street
London, SE19SG - United Kingdom
Printed in Croatia

British Library Cataloguing-in-Publication Data
A catalogue record for this book is available from the British Library

Additional hard copies can be obtained from orders@intechopen.com

Advanced Sorption Process Applications
Edited by Serpil Edebali
p. cm.
Print ISBN 978-1-78984-818-2
Online ISBN 978-1-78984-819-9

For EU product safety concerns:
IN TECH d.o.o., Prolaz Marije Krucifikse Kozulić 3, 51000 Rijeka, Croatia, info@intechopen.com or visit our website at intechopen.com.

Meet the editor

Associate Professor Dr. Serpil Edebali was born in Konya, Turkey, in 1977. She received her BS degree at the Food Engineering Department of METU in Ankara. Her MSc and PhD degrees were received at the Chemical Engineering Department of Selcuk University in Konya in 2004 and 2010, respectively. From 2002 to 2013, she was a research assistant at the Chemical Engineering Department of Selcuk University. Since 2013, she has been assistant professor and is now working as associate professor doctor in the chemical technologies section of the same department. She gives lectures in the subjects of heat transfer and fluid mechanics. Her research interests include wastewater treatment, removal of heavy metals and dyes, synthesis of nanoparticles and applications, separation processes, especially sorption with resins and membranes, as well as membrane-adsorption hybrid systems. She has duties as a researcher in various COST and FP7, Erasmus+ as well as H2020 projects.

Contents

Preface

At the beginning of the twenty-first century, separation processes presented a comprehensive application of the major operations performed by various industries, such as chemical, food, environmental, and biotechnology. Sorption, one of the preferred separation processes because of its effectiveness at different interfaces, has caught the attention of many scientists. This book is aimed at gaining a general knowledge of sorption and a number of extremely important applications, as well as recognizing its functions and paramount importance in chemical and biochemical plants, including environmental treatment. Moreover, progress in the phenomenon is highlighted in this book.

To help provide instruction in the important sorption processes, we have chosen authors who have extensive industrial and academic experience in closing the gap between theory and practice. Crucial progress in the theoretical information section of sorption has been achieved, mainly through the development of new techniques that examine the usage of various sorbents, including nanomaterials for the removal of various pollutants. We have subdivided the book into various sections, one of which is focused on applications of the sorption process, which presents real results of the recent studies and gives a source of up-to-date literature. The relationship between the sorption process and isotherm and kinetics modeling is analyzed in another chapter. This book will be a reference book for those who are interested in sorption techniques from various industries.

For their frankness and diligence in reviewing the proposed chapters, I would like to thank all authors who have contributed to this book. We have incorporated many of their recommendations and the book is much improved as a consequence.

Special thanks goes to Dolores Kuzelj (Author Service Manager) for her dedicated support in the reviewing process and her suggestions for further improvement. Finally, all thanks to IntechOpen for publishing this book.

Dr. Serpil Edebalı
Chemical Engineering Department,
Faculty of Engineering, Selcuk University,
Konya, Turkey

Section 1

Natural Sorbents

Chapter 1

Biopolymer-Based Materials from Polysaccharides: Properties, Processing, Characterization and Sorption Applications

Rohan S. Dassanayake, Sanjit Acharya and Noureddine Abidi

Abstract

Biopolymers are polymeric materials derived from biological sources. Due to their renewability, abundance, biodegradability and other unique properties such as high adsorption capabilities and ease of functionalization they have been investigated for several industrial applications including sorption. Polysaccharides especially cellulose, chitin and chitosan are important biopolymers because of their high abundance, wide distribution and low cost of production. This chapter provides an overview of properties, common processing methods, and material characterization of three commonly studied biopolymers namely cellulose, chitin and chitosan. It provides a thorough review on recent developments on utilization of cellulose, chitin, and chitosan-based materials for various sorption applications. Specifically, their application and efficiency in organic dye removal, heavy metals removal, oil and solvent spillage cleanup, and CO_2 adsorption are presented and discussed.

Keywords: cellulose, aerogels, chitin, functionalized cellulose aerogels, biopolymer-based materials

1. Background

1.1 Biopolymers

As the name suggests, biopolymers are polymers synthesized by living organisms. Therefore, they are polymeric biomolecules i.e. long chain biomolecules comprised of covalently linked repeating monomeric units [1]. Living organisms (plants, animals, bacteria, fungi and yeast) synthesize a wide range of biopolymers such as deoxyribonucleic acid (DNA), ribonucleic acid (RNA), proteins, cellulose, chitin, starch, etc. Biopolymers carry out countless number of vital functions, such as storage of energy, preservation and transmittance of genetic information, and cellular construction, in vivo. DNA and RNA are the hereditary materials for the storage and passage of the genetic information in all living organisms, and thus making perpetuation of life possible in the planet. Proteins not only catalyze reactions (e.g. enzymes) and take part in cell signaling (e.g. hemoglobin) but also provide structural support (e.g. collagen). Cellulose is the major structural component of the plant cell walls.

There are three main classes of biopolymers owing to their universal occurrence and abundance: (i) polynucleotides, (ii) polypeptides/poly amino acids, and (iii) polysaccharides. Polynucleotides (DNA and RNA) are long polymers composed of 13 or more nucleotide monomers [1]. Polypeptides are the short polymers comprised of amino acids as monomeric units and amide bonds link the monomeric units together [2]. Polysaccharides are composed of monomeric sugars linked together by O-glycosidic linkages. Among hundreds identified polysaccharides, cellulose, starch, chitin, chitosan are some of the important examples [3–5]. In contrast to the often simpler and more random structure of the synthetic polymers, biopolymers, in their natural states, are complex molecular assemblies that assume defined 3D structure and shapes, often known as hierarchical levels of structure.

Since biopolymers possess material properties suitable for various industrial and medical applications, they have garnered a great deal of interest both in academia and industry. Primarily, the interest and increasing trend of application of biopolymers as commodity products stems from their renewability, biodegradability and, often, their biocompatibility. Even though traditional plastics (petroleum based polymer products) are still indispensable in our daily life, there is a growing concern about their environmental impacts as they are non-biodegradable and as a consequent, accumulate in the environment. Furthermore, petroleum is not a renewable resource and its reserve is dwindling at a fast pace. Thus, biopolymers derived from renewable resources possess competitive advantage over synthetic non-renewable polymers. However, the cost of production of biopolymers and biopolymer based commodity products is of prime importance if they are to compete with oil-based synthetic polymers in the market.

Exploiting biopolymers, which are abundant and can be obtained from source material at lower cost, as a main component of biopolymeric products can be a smart strategy in the first place to mitigate the cost related issues. To that end, polysaccharides, the biopolymers comprised of monosaccharides (sugars) linked together by O-glycosidic linkages are of obvious choice. Both the most abundant natural resource, cellulose, and the second most abundant biopolymer, chitin, are polysaccharides [3, 6]. Additionally, polysaccharides are widely distributed in nature as they can be derived from plants, animals and microorganisms. Furthermore, variation in physicochemical properties, such as mechanical properties, solubility, viscosity, gelling potential, surface and interfacial properties, governed by monosaccharide composition, chain length (degree of polymerization), linkage types and patterns, provide polysaccharides versatility in preparation of materials with diverse applications. In fact, polysaccharides based materials in different forms including fibers, films, food casing, membranes, hydrogels, aerogels and sponges, with applications in several important commercial areas such as food, pharmaceuticals, biomedical, electronics, and adsorption have been developed [7–13]. Thus, polysaccharide-based biopolymers are promising candidates in the preparation of materials that can meet the much coveted dual requirements of environmental friendliness and economic sustainability. In this book chapter we focus on three important polysaccharide based biopolymers: cellulose, chitin, and chitosan.

1.2 Cellulose

Cellulose, the most abundant renewable resource in the planet with an annual yield of 1.5×10^{12} tons, is the major structural component in plant cell walls. Besides plants, some species of bacteria, algae also biosynthesize cellulose. Tunicates are the only known animals capable of biosynthesizing cellulose [3, 14]. Because of its abundance, cellulose can serve as a virtually inexhaustible source of raw material in production of sustainable bioproducts, the so called "green products" [3]. It is a linear homopolymer of β-D glucose monomers covalently linked together by 1→4 glycosidic bonds i.e. the ringed glucose monomers are joined together via polycondensation reaction between hydroxyl groups at C1 of a glucose unit and C4 of the neighboring glucose unit. Since, every glucose unit is rotated 180° with respect to its neighbors along the fiber axis, dimer of glucose, called cellobiose, is considered as a repeating unit of cellulose polymer (**Figure 1a**) [15]. However, the convention of considering cellobiose as a repeating unit has been challenged and a recent review in this regard makes a cogent argument that glucose is the repeating unit rather than the cellobiose [16]. Cellulose is a long chain molecule and the degree of polymerization (DP) differs with source and the treatments employed during its extraction. The DP of cellulose is reported to be as high as 10,000 and 15,000 in wood and cotton fibers [14, 17].

Cellulose in nature is not found as an isolated molecule, but rather possesses an intriguing multi-level assembly, popularly known as hierarchical structure [18]. It is usually agreed that, during biosynthesis, approximately 36 individual cellulose molecules are assembled together to form elementary fibrils, which have dimensions in nano-scale (a square cross section, 3–5 nm in size). Elementary fibrils further undergo packing into a larger entity called microfibrils with a cross section of ~20 nm × 8 nm. These microfibrils further assemble into macro sized cellulose fibers (e.g. wood fiber, cotton fiber) [14, 15]. However, specific packing of cellulose may be different based on the source.

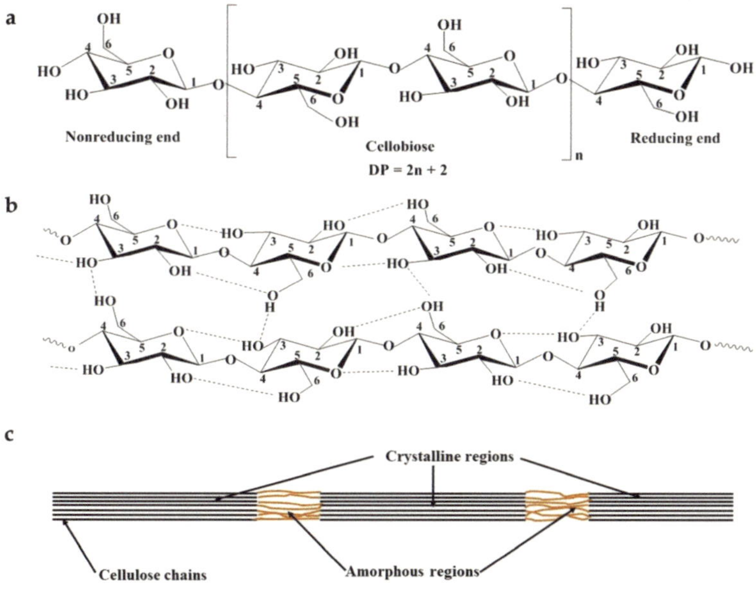

Figure 1.
Chemical structure and hydrogen bonding in cellulose: (a) chemical structure of cellulose, (b) hydrogen bonding in cellulose, and (c) schematic of cellulose microfibrils showing crystalline and amorphous structure.

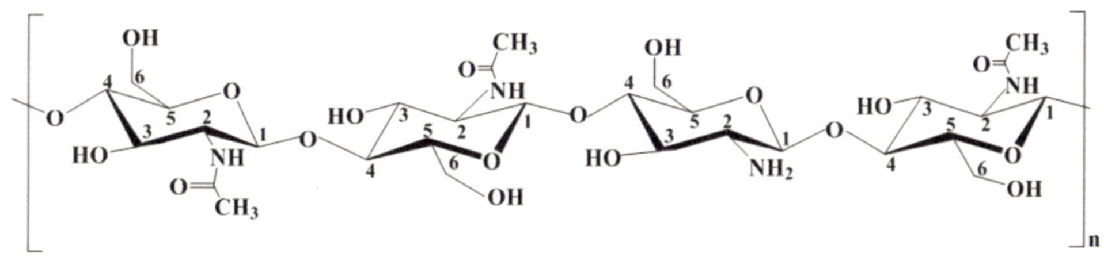

Figure 2.
Chemical structure chitin.

Prevalence of hydroxyl groups in cellulose (three hydroxyl groups per glucose monomer) results in extensive intra-and interchain hydrogen bonding in cellulose (**Figure 1b**). Intra-and interchain hydrogen bonding and the stacking interactions mainly originating from Van der Waals forces facilitate highly ordered arrangement (crystalline regions) of cellulose fibrils although there are also amorphous regions present (where arrangement of cellulose chains is disordered) as depicted in **Figure 1c** [19]. In general, cellulose, regardless of the source, is highly crystalline and high molecular weight biopolymer and is, usually, a fibrous, tough and hydrophilic but insoluble in water and other common solvents [20]. However, important properties such as DP, crystallinity and fibrillar organization are dependent on the source. For example, cellulose from cotton fiber is more crystalline (~80% crystallinity) as compared to wood cellulose (~50% crystallinity). Similarly, quite different from plant derived cellulose, bacterial cellulose has ultrafine nanofibrous network structure, high hydrophilicity, and moldability [21].

1.3 Chitin

Chitin is another important polysaccharide and the most abundant biopolymer on the planet after cellulose. It is a structural biopolymer, contributing in strength and reinforcement, present in the exoskeleton of arthropods (i.e., in the shells of shrimps and crabs), and the cuticle of insects. It is also found in the cell walls of fungi, yeast and other organisms in lower plant and animal kingdoms [4]. Structurally, it is similar to cellulose except the presence of acetamide or amine functional groups at C2 position of the glucose monomer unit. Therefore, chitin is a linear polymer of 2-acetamido-2-deoxy-D-glucopyranose linked together by $\beta(1\rightarrow4)$ glycosidic bond [22] (**Figure 2**). The degree of acetylation (DA) which is the share of nitrogen sites occupied by acetyl group is typically ~90% in chitin after extraction and purification from source material since alkali treatment applied for the removal of protein also lowers DA [23].

Similar to cellulose, native chitin is arranged in highly crystalline microfibrils because of the extensive hydrogen bonding among the polymer chains. It is tough, hydrophobic, and insoluble in water and other common organic solvents. Chitin is even more recalcitrant to solubilization and consequently harder to process than cellulose [6].

1.4 Chitosan

Chitosan is the most important partially deacetylated derivative of chitin with degree of acetylation (DA) <50%. Typically chitin has DA <35% [24]. Chitin can be deacetylated by treating it with concentrated alkali at elevated temperature. Enzymatic hydrolysis in the presence of chitin deacetylase can also be employed for deacetylation purpose [21]. Therefore, chitosan can be represented by the same chemical structure as chitin. While chitin is insoluble in common solvents, chitosan can be dissolved in weak acidic solutions (aqueous solutions with pH < 6.5). It is soluble in hydrochloric acid and aqueous organic acids such as formic, acetic, oxalic and lactic acids. Protonation of the free glucosamine in acidic condition facilitates chitosan solubility [25, 26]. Chitin and chitosan are reported to have inherent antimicrobial properties.

2. Processing of cellulose, chitin and chitosan

2.1 Derivatization of cellulose, chitin and chitosan

Because of the abundance of hydroxyl (OH) groups present in cellulose, it can be derivatized through esterification, etherification, and other reactions to produce cellulose derivatives such as acetylated (e.g. cellulose acetate), carboxy methylated (e.g. carboxymethyl cellulose), and phosphorylated (e.g. cellulose diphenyl-phosphate) and cationized cellulose [27–30]. Cellulose acetate (CA) is one of the most important commercially available cellulose derivatives with many applications such as fiber, film, and osmotic membrane. CA is a cellulose ester in which hydroxyl groups in cellulose are esterified with acetate groups to varying degrees [31]. Similarly, chitin and chitosan can be derivatized through O- and N-substitution to obtain carboxyalkyl derivatives such as O-carboxymethyl chitin, O-carboxymethyl chitosan and N,N-dicarboxymethyl chitosan [21, 24]. Derivatization not only eases the processing of these biopolymers since derivatives can be solubilized in common solvents and even in water based on degree and type of substitution [21, 29, 32], but also imparts some novel chemical and physical properties to cellulose. For instance, phosphorylation of cellulose can impart enhanced flame retardancy and bioactivity [21, 32]. Similarly, carboxyalkylation of chitosan enhances its affinity towards metal ions and has application in metal ions collection [24].

2.2 Dissolution of cellulose, chitin and chitosan

Due to their high molecular weight and complex biopolymeric network (extended hydrogen bonding and crystallinity), processing of cellulose and chitin into desirable forms, such as films, membranes, fibers, and gels, via dissolution is generally considered as nontrivial task as they are insoluble in water and other common organic solvents [13, 33]. Therefore, although studied for a long time, cellulose and chitin dissolution in cost effective and environmentally friendly manner remains an active area of research. A number of solvents/solvent systems have been developed over the course of time. Derivatizing solvent (carbon disulfide/aqueous sodium hydroxide used in "Viscose" process), and other non-derivatizing solvents such as aqueous alkali solution (e.g. aqueous solution of 7% sodium hydroxide/12% urea), *N,N*-dimethylacetamide/lithium chloride (DMAC/LiCl), *N*-methylmorpholine-*N*-oxide (NNMO) (used in "Lyocell" process), and ionic liquids (ILs) have been some of the important cellulose solvents [34–38]. Aqueous sodium hydroxide/urea solution, DMAc/LiCl and ILs such as 1-allyl-3-methylimidzolium bromide, 1-butyl-3-methylimidzolium chloride and 1-butyl-3-methylimidazolium acetate have been reported to be capable of dissolving chitin [33]. Although they are yet to find significant industrial breakthrough, ionic liquids are thought to have advantages over other solvents due to their chemical and thermal stability, tunability, non-flammability, negligible low vapor pressure, high efficiency in dissolution of both cellulose and chitin. Regenerated cellulose products such as fibers, film, aerogels and beads have been prepared from biopolymer solutions in different solvents [39–42].

2.3 Aerogels

Aerogels are broadly defined as sol–gel derived materials, which are highly porous, low density and predominantly mesoporous (pore diameter: 2–50 nm) [43]. They are derived from hydrogels followed by the removal of liquid component using super critical point drying or freeze-drying to avoid damage to the original gel micro-structure prepared from biopolymers solution. Because of their very low density, high porosity and high specific surface area, aerogels have potential applications in separation, adsorption, catalyst, photo-electricity, and biomedicine [13]. High surface area aerogels from cellulose (400–500 $m^2\ g^{-1}$) [44], chitin (560 $m^2\ g^{-1}$) [45], and chitosan (248 $m^2\ g^{-1}$) [46] have been prepared.

2.4 Nanofibrillation/nanoparticles (nanocellulose, nanochitin, nanochitosan)

The term "nanocellulose"/"nanochitin"/"nanochitosan" encompasses various materials derived from respective biopolymers, which possess at least one dimension in the nanometer range [14, 47]. Nanoparticles from cellulose and chitin are usually prepared by destructing the native hierarchical structure of these biopolymers [48]. Cellulose and chitin nanofibers, in general, are obtained by subjecting purified cellulose and chitin substrates to multiple mechanical shearing actions, which disintegrate the native microfibril structure and release enmeshed individual or bundle of fibrils. Chitosan nanofibers are typically produced by electrospinning [49].

Highly crystalline elongated rod like (or needle-like) nanoparticles called nanocrystals can be obtained when cellulose and chitin substrates are subjected to a strong acid hydrolysis treatment due to preferential dissolution of amorphous domains. For acid hydrolysis to produce nanocrystals, sulfuric acid and hydrochloric acid are usually employed [50, 51]. Biopolymer nanoparticles can be directly processed into films and aerogels by drying from the suspension or they can be utilized as reinforcement agents in other polymer matrices.

3. Characterization of biopolymers

The most widely used characterization techniques of biopolymers include Fourier transform infrared spectroscopy (FTIR), X-ray diffraction (XRD), thermogravimetric analysis (TGA) and scanning electron microscopy (SEM). In this section, we discuss the characterization of polysaccharides: cellulose, chitin, and chitosan using FTIR, XRD, TGA and SEM.

3.1 Materials

Microcrystalline (MCC) was provided by FMC Biopolymer (Newark, DE). Chitosan (75–85% deacetylated) and chitin from shrimp shells were used as received from Sigma-Aldrich (Saint Louis, MO).

3.2 Instrumentation

3.2.1 X-ray diffraction (XRD) analysis

The wide-angle powder X-ray diffraction patterns of the samples were recorded on a SmartLab XRD system (Rigaku Corporation, ModelHD2711N) with CuK$_\alpha$ radiation (λ = 1.541867 Å). The accelerating voltage and tube current used were 40 kV and 44 mA, respectively. A continuous scanning was performed at a scan speed of 2°/min and the 2θ ranged from 10° to 60°.

3.2.2 Thermogravimetric analysis (TGA)

Thermogravimetric (TG) analyses of cellulose, chitin and chitosan samples were obtained using a PerkinElmer Pyris1TGA instrument (PerkinElmer, Waltham, MA) furnished with a 20-sample auto-sampler. TG profiles were recorded in an inert nitrogen atmosphere (20 ml/min) from 37 to 600°C with a constant a heating rate of 10°C/min using a high-resolution mode. All data were analyzed using Pyris Data Analysis software.

3.2.3 Scanning electron microscopy (SEM)

The morphology of cellulose, chitin and chitosan samples was studied on a Hitachi S-4700 field emission scanning electron microscope (TM-100, Hitachi, Japan) with an accelerating voltage of 15 kV.

3.2.4 Fourier transform infrared (FTIR) spectroscopy

FTIR spectra of the cellulose, chitin and chitosan samples were recorded using a PerkinElmer Spectrum-400 FTIR spectrometer equipped with a universal attenuated total reflectance (UATR) accessory (PerkinElmer, Waltham, MA). All FTIR spectra were collected at a spectral resolution of 4 cm^{-1}, with 32 co-added scans in the wavenumber range of 4000 to 650 cm^{-1}. The spectra were analyzed using PerkinElmer Spectrum software.

All samples were conditioned in an environmentally-controlled laboratory maintained at a relative humidity of 65 $\pm$ 2% and temperature of 21 $\pm$ 1°C for at least 48 h prior to their characterization.

4. Characterization of cellulose, chitin and chitosan

4.1 X-ray diffraction (XRD) analysis

Wide-angle X-ray diffraction measurements were collected for raw cellulose, chitin, and chitosan powder samples. **Figure 3** shows the X-ray diffraction curves for cellulose, chitin, and chitosan. The XRD pattern of cellulose exhibits five major diffraction peaks at 14.9 (101), 15.8 (10$\bar{1}$), 21.8 (021) 22.5 (002) and 34.6 (004) which are in agreement with the literature values reported for cellulose I_β [52]. The XRD patterns of chitin showed two strong reflections at 9.2° (020) and 19.1° (110) and minor reflections at 12.6° (021), 22.9° (130) and 26.2° (013) corresponding to α-chitin [53]. Chitosan shows two distinct peaks at 10.67 (020) and 19.92° (110) [54]. Our results suggest that chitin has the highest crystallinity as compared

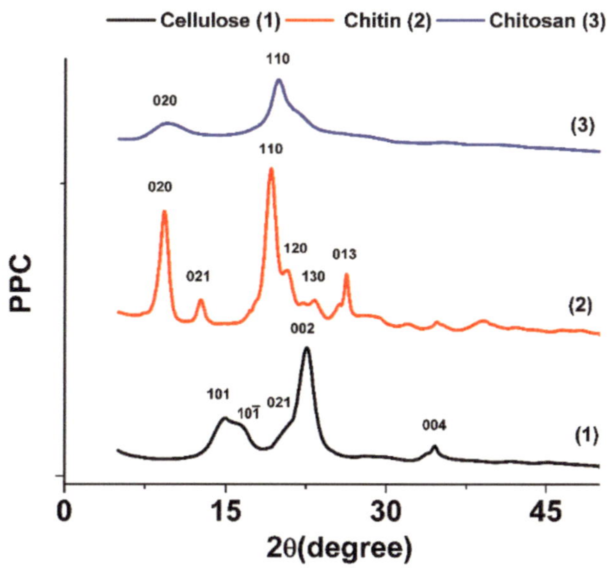

Figure 3.
Wide angle X-ray diffraction curves for cellulose, chitin and chitosan powder.

to cellulose and chitosan. The low crystallinity of chitosan is the result of the deacetylation process.

4.2 FTIR characterization

FTIR studies were performed on cellulose, chitin and chitosan powders. **Figure 4** shows the UATR-FTIR spectra of cellulose, chitin, and chitosan collected in the wavenumber range of 4000–600 cm^{-1}. The peaks of cellulose at 3335 and 3234 cm^{-1} are attributed to the O-H stretching vibrations and the extensive inter- and intra-molecular hydrogen bonding network [55]. For chitin, the vibrations attributed to O-H and N-H stretching are visible at 3434 and 3256 cm^{-1} and for chitosan those peaks appear at 3434 and 3256 cm^{-1} [56, 57]. The bands in the region of 3000–2800 cm^{-1} correspond to the -CH_2 symmetrical and asymmetrical stretching vibrations of polysaccharides [55–57]. Chitin exhibits the doublet amide I band at 1658 and 1619 cm^{-1} representing the presence of H-bonding in a C=O group with the NH group of the adjacent chain and the O-H group of the inter-chain [58]. The amide II band of chitin and chitosan is located at 1553 and 1586 cm^{-1}, respectively [56, 57]. The peak at 1307 cm^{-1} corresponds to the amide III band (in-plane mode of $CONH_2$ group) of chitin. Respective, amide I and C-N stretching of amide III of chitosan appear at around 1647 and 1318 cm^{-1}. The vibrations at ~1152 and ~1110 cm^{-1} are attributed to the anti-symmetrical bridge of C-O-C stretching and anti-symmetric in-plane stretching. The peaks at ~1060 and ~1025 cm^{-1} correspond to the C-O stretching of the polysaccharide [55–57]. The band at ~895 cm^{-1} is assigned to β-linkage of the polysaccharide [55–57].

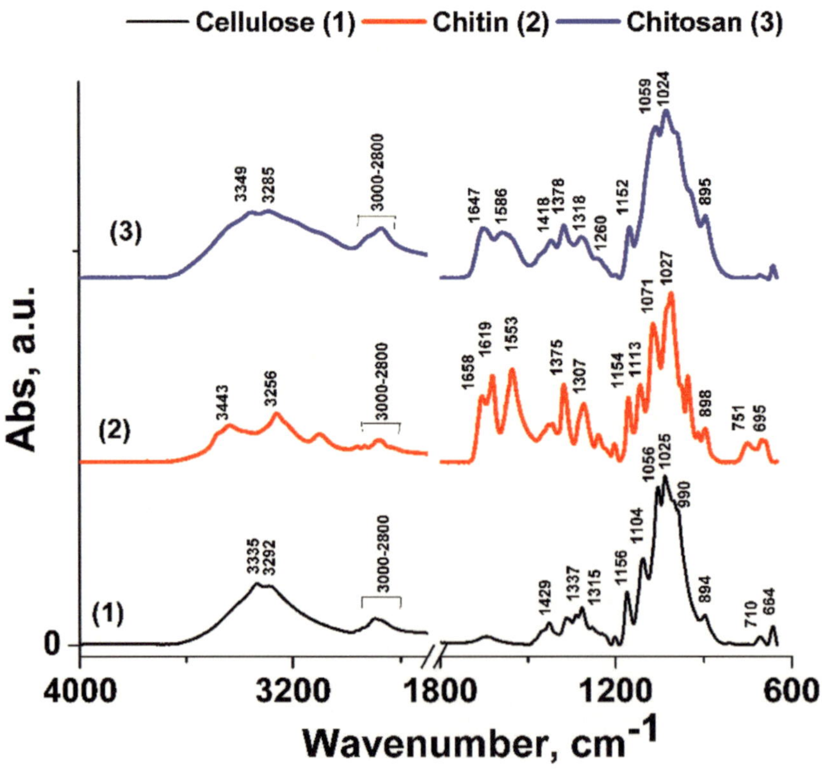

Figure 4.
FTIR spectra of cellulose, chitin, and chitosan in the wavenumber region of 4000–600 cm^{-1}.

4.3 TG and DTG analysis

High-resolution thermogravimetry (TG) profiles recorded in flowing nitrogen (N_2) were used to investigate the thermal stability of cellulose, chitin and chitosan, see **Figure 5(a)**. All samples display two main weight loss regions at 35–150 and 250–450°C. In the first region, approximately, 5.1–9.0% of the weight loss was observed, which is attributed to the removal of physically adsorbed water. The second region represents the largest weight loss of 90.4, 86.6 and 63.4% for cellulose (1), chitin (2) and chitosan (3), respectively, corresponding to the degradation of polysaccharide structure of the biopolymer. The differential thermogravimetric (DTG) profiles generated from TG data are shown in **Figure 5(b)**. The DTG curves for cellulose, chitin, and chitosan exhibit the main decomposition peaks at 371, 391 and 300°C, respectively. These results suggest that among three biopolymers, chitin has the highest thermal stability while chitosan has the least thermal stability.

4.4 SEM studies

The morphological characteristic of cellulose, chitin and chitosan was investigated and **Figure 6** shows the comparison of typical surface morphologies for cellulose, chitin and chitosan. As it can be seen, all samples exhibited more irregular, flat, rough nanofiber surface with no porosity.

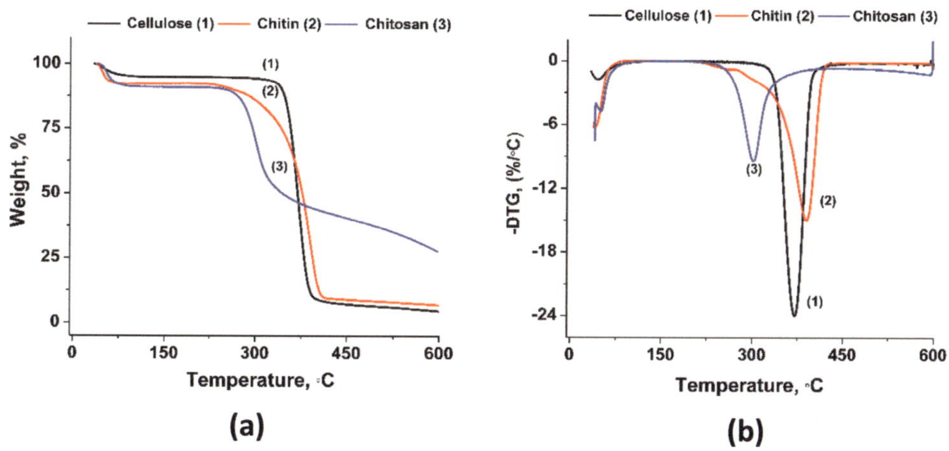

Figure 5.
(a) TG and (b) DTG profiles of cellulose, chitin and chitosan.

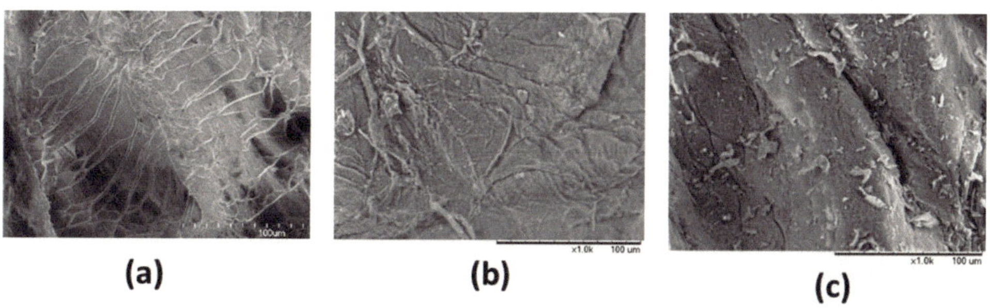

Figure 6.
SEM images of (a) microcrystalline cellulose, (b) chitin and (c) chitosan at 100 um magnification.

5. Applications of biopolymer-based materials in sorption

With the rapid industrialization, more environmental concerns (such as industrial discharge of organic dyes and heavy metal ions, oil and solvent spillage, and emission of greenhouse gases including carbon dioxide (CO_2)) have emerged as a critical worldwide issue, imposing detrimental damages to the environment, economy and human health. Recently, biopolymer-based materials being sustainable, readily available, biodegradable, economical, and environmentally-friendly, they are widely investigated both in industry and research as a green approach in remediating those issues. In this section, we discuss the use of cellulose-, chitin-, and chitosan-based materials as sorbents for four different applications including removal of organic dyes and heavy metals, oil and solvent spillage cleanup and CO_2 adsorption.

5.1 Dye removal

Industries including textile, leather, paper, printing, and paint utilize large amounts of water for processing have the potential to contaminate waterways through the discharge of organic dyes into natural water resources and water treatment systems. Over 10,000 different textile dyes with an estimated annual production of over 7×10^5 tons are available commercially [59, 60]. Most of these organic dyes are excessively used and 10–20% is directly discharged as aqueous effluents into different water bodies [61]. Most problematic ones are the brightly colored, water-soluble reactive and acid dyes as these dyes cannot be removed through conventional treatment systems. Discharging of organic dyes into water bodies has raised acute and chronic concerns to the ecosystems and human health. For example, the release of those organic dyes can lead to eutrophication, non-esthetic pollution and imbalance in the aquatic biological systems and also causes chronic toxicity, carcinogenicity and neurotoxicity towards humans and animals. Commonly applied technologies for removing organic dyes from wastewater include chemical precipitation and adsorption [62, 63], electrochemical oxidation and reduction [64], aerobic and anaerobic treatment [65, 66], coagulation and flocculation [67, 68], membrane separation [69], ultra-filtration [70], H_2O_2/ultraviolet (UV) and photocatalysis [71–73], ion exchange [74], sonochemical degradation [71], Fenton and heterogeneous Fenton-like catalysis [75–77], electrolysis [78], and advanced oxidation processes (AOPs) [56, 79]. At present, there is a critical need for research that enables new means of inexpensive, reusable, environmentally and energetically sustainable wastewater management systems for wastewater treatment. Recently, biopolymer-based materials have attracted interests as emerging alternative techniques for the removal of organic dyes from industrial effluents due to their relative abundance, eco-friendly, high thermal and mechanical stability, low cost, nontoxicity, easy functionalization, and excellent sorption capacities. Here, we report the roles of biopolymer-based materials mainly derived from cellulose, chitin, and chitosan as sorption materials for the removal of organic dyes from industrial wastewater.

Liu et al. investigated the porous cellulose-based bioadsorbent for the removal of anionic dye acid blue 93 (AB93) and cationic dye methylene blue (MB) from single and binary dye solutions by grafting acrylic acid and acrylamide onto cellulose surface [80]. They reported maximum sorption capacity of 1372 mg/g at an initial absorbent concentration of 2500 mg/L for both of AB93 and MB with a reusability up to three consecutive cycles. The adsorption process followed the pseudo-second-order kinetic model and the Freundlich isotherm model. Jiang and coworkers prepared ultralight aerogels from cellulose nanofibrils (CNFs) derived from rice straw [81]. These ultra-lightweight aerogels have been studied as adsorbents for the removal of cationic dye malachite green (MG) dye from aqueous media. The highest adsorption capacity of MG was found to be 212.7 mg/g. They also reported a 100% MG removal via four consecutive adsorptions at a 1:5 mg/mL aerogel/MG ratio and 10 mg/L dye concentration and 92% MG adsorption in a single batch at 10:5 mg/mL aerogel/MG ratio and 100 mg/L dye concentration. The adsorption process obeyed a pseudo-second-order kinetic model and monolayer Langmuir adsorption isotherm. The dye desorption was also achieved by increasing ionic strength, leading to the recovery of both dye and CNF aerogel. Ruan and coworkers prepared porous 2,3-dialdehyde cellulose-chitosan (DAC-CS) beads and their applications in the removal of Congo red dye [82]. DAC-CS beads showed 100% dye removal at a capacity of 200 mg/g at pH. 2 in 100 g/L dye solution. About 93% of the adsorbed Congo red in DAC-CS beads could be desorbed at pH. 12 NaOH solution. A similar trend was observed in the second cycle with the adsorption and desorption of 100 and 89%, respectively. Cross-linked quaternized cellulose nanofibrils (Q-CNFs) have been investigated as adsorbents for removal of three different anionic dyes by Maatar and Boufi [83]. They reported the adsorption capacity of approximately 160, 230, and 560 mg/g for red 180, blue CR19, and orange 142 dye, respectively. The authors also confirmed the main contribution of electrostatic interactions between positive sites and the CNF surface and dye sulfonate groups by zeta potential measurements. The regeneration of the Q-CNF sorbent could be achieved by extraction with KCl solution in ethanol-water mixture.

Kinetics of the adsorption of reactive yellow 2 (RY2) and reactive black 5 (RB5) by chitin was investigated by Akkaya et al. [84]. They concluded that the adsorption of RY2 by chitin from aqueous solution is favorable at low temperature and pH and the adsorption of RB5 by chitin is favorable at high temperature and pH. Cao and coworkers studied the adsorption isotherms and kinetic analysis of methylene blue (MB) onto porous chitin sorbents (PChs) with different content of chitin, ranging from 0.9 to 3.5% [85]. The maximum adsorption capacity of 384 mg/g with 79.8% MB removal was observed for PChs containing 3.5% chitin. The same adsorption capacity with 65% removal ratio was observed after six repetitive cycles. The adsorption equilibrium followed Freundlich isotherm model. Xu and coworkers reported the synthesis and efficient removal of MB by nanocomposite microspheres fabricated from chitin and clay [86]. These chitin/clay microspheres (CCMs) showed a maximum adsorption capability of 156.7 mg/g. A dye removal of 99.99% was observed within 20 at a low MB concentration of 10 mg/g. They also demonstrated a stability over five adsorption/desorption cycles for CCMs. Their kinetic data fitted well with the pseudo-second-order and monolayer Langmuir isotherm models. Gopi et al. prepared a chitin nanowhisker (*ChNW*)-functionalized polyvinylidene fluoride (PVDF) membrane using electrospinning technique [87]. They reported a maximum adsorption capacity of 72.6 mg/g and 88.9% removal efficiency of indigo carmine (IC) dye with their *PVDF/ChNW* (15%:1% wt.) composite membrane as compared to neat PVDF.

Xu and coworkers synthesized poly(2-acrylamido-2-methylpropane sulfonic acid grafted magnetic chitosan microspheres (PMCMs) for MB adsorption [88]. They investigated the effects of initial solution pH (1.0–10.0), temperature (30–50°C), contact time (0–660 min) and initial concentration (50–1600 mg/L) on dye adsorption kinetics. At pH. 9, PMCMs exhibited maximum MB sorption capacity of 1000, 1250 and 1428 mg/g at 30, 40 and 50°C, respectively. They also showed that the adsorption capacity increased with increasing of the initial solution pH and temperature. The adsorption kinetic and adsorption equilibrium data followed the pseudo-second-order kinetic model and monolayer Langmuir isotherm model, respectively. PMCMs could also be separated under external magnetic field and regenerated under acidic conditions. Naseeruteen et al. exhibited the removal of Malachite Green (MG) by mesoporous chitosan ionic liquid beads prepared from chitosan and 1-butyl-3-methylimidazolium based ionic liquids [89]. The effects of initial pH, adsorbent dosage, agitation time and initial MG concentration have also been investigated. The optimum conditions reported were pH 4.0, 0.008 g of adsorbent dosage and 20 min of agitation time. The data obtained at pH 4 fitted well to a pseudo-second order kinetic model and followed the Langmuir model. The maximum adsorption capacity obtained for two chitosan beads prepared from 1-butyl-3-methylimidazolium acetate and 1-butyl-3-methylimidazolium was 8.07 and 0.24 mg/g respectively.

5.2 Removal of heavy metals

The surge in energy production and the increasing use of heavy metals in various industrial processes have led to the generation of large quantities of industrial waste containing heavy metals. Therefore, heavy metals are released into the environment including air, water, soil, and biosphere in excessive amounts on daily basis. Heavy metal ions are mainly elements from the fourth period of the periodic table including chromium (Cr), arsenic (As), cobalt (Co), copper (Cu), nickel (Ni), zinc (Zn), lead (Pb), and mercury (Hg) and have a high atomic weight and a density at least 5 times higher than that of water [90]. Heavy metals are typically toxic, persistent, non-biodegradable and bioaccumulative. Due to the high solubility, heavy metals are easily absorbed by plants and aquatic species and subsequently enter the food chains and then the human body. The presence of high levels of heavy metals in the human body may cause various health effects including skin irritations, stomach cramps, vomiting, multiple organ damage, birth defects, nerve system damage and development of autoimmunity. Most commonly used techniques for the removal of heavy metals include membrane filtration, adsorption, coagulation, chemical precipitation, ion-exchange, electrochemical, biological treatments and advanced oxidation processes [91]. However, adsorption technologies are widely considered as cheap and energy efficient solutions for removal of heavy metals. Biopolymers including cellulose, chitin, and chitosan have been widely investigated as sorbents for remediation of heavy metals due to their excellent sorption properties and environmental benignity [92–94]. Here we present, recently reported applications of biopolymer-based materials for the removal of heavy metals.

d'Halluin et al. prepared a chemically modified cellulose filter paper with ethylenediaminetetraacetic acid (EDTA) for heavy metal remediation in water [95]. Cellulose-EDTA composite showed 90–95% removal efficiency for various heavy metal ions including Ag(I), Pb(II), Cd(II), Ni(II), Zn(II), Sn(II), and Cu(II). They also investigated the adsorption kinetics using Langmuir, Freundlich, and Temkin isotherms and the recyclability of their material. Fakhre and coworkers synthesized a supramolecular polysaccharide composite material from cellulose (CEL) and dibenzo-18-crown 6 (DB18C6) using ceric ammonium nitrate as initiator [96]. They studied the removal of five different heavy metal ions including Cd^{2+}, Zn^{2+}, Ni^{2+}, Pb^{2+} and Cu^{2+} by [CEL + DB18C6] composites. They reported adsorption capacities and percentage of metals removal of (197 mg/g, 98%), (180 mg/g, 94%), (186 mg/g, 93%), (194 mg/g, 97%), and (192 mg/g, 96%) for Cd^{2+}, Zn^{2+}, Ni^{2+}, Pb^{2+} and Cu^{2+}, respectively. The regeneration of the material is dependent on the type and concentration of regenerating solution (NH_4Cl, HNO_3, NaCl and $CaCl_2$). Shao et al. studied the removal of Cu^{2+} ions by a porous waste paper (WP)-chitosan adsorbent (CSA) [97]. WP-CSA exhibited adsorption capacity of 156.3 mg/g for Cu^{2+} and

98.3% desorption efficiency in 0.1 M H_2SO_4 solution. This adsorption equilibrium followed the pseudo-second order model and the Langmuir monolayer model.

Duan and coworkers presented the fabrication of lignin/chitin films from a binary solvent system composed of the ionic liquid 1-butyl-3-methylimidazolium acetate and γ-valerolactone and their application as an adsorbent for Fe(III) and Cu (II) cation uptake from aqueous solutions [98]. Lignin/chitin film showed the adsorption capacity of 84 and 22 wt% for Fe(III) and Cu(II) cations within 48 h. During the regeneration process up to 12 and 46 wt%, respectively, could be desorbed. The adsorption isotherms for both Fe(III) and Cu(II) obeyed the Langmuir model. Ethylenediaminetetraacetic acid (EDTA) modified β-cyclodextrin/ chitosan (CDCS-EDTA) composite was prepared and applied for the removal of Pb (II) and anionic dye acid red 73 (AR) by Wu and coworkers [99]. They also investigated the effects of adsorbent dose, contact time and pH value on the CDCS-EDTA adsorption efficiency. The maximum adsorption capacities reported were 114.8 and 754.6 mg g^{-1} for Pb(II) and AR under optimal conditions with adsorption efficiently of 93.4 and 92.1%, respectively. Both adsorption processes followed the pseudo-second-order adsorption kinetic model and the Langmuir isotherm model. Kwok et al. studied the sorption of anionic species arsenite, As(III) ions and arsenate, AS (V) ions onto chitosan and nanochitosan [100]. The equilibrium adsorption capacity of chitosan and nanochitosan for As(III) and As(V) ions were (0.5 and 8 mg/g) and (6.1 and 13 mg/g), respectively.

5.3 Oil and solvent spillage cleanup

An oil or a solvent spill is the accidental or intentional discharge of petroleum hydrocarbons into the environment, especially the aquatic ecosystem. Oil and solvent spills have become a serious environmental problem due to the increasing use, exploration, production and transportation of oils and solvents worldwide by various petrochemical and chemical industries. Those spillages lead to the pollution of aquatic ecosystems and natural water resources including underground water streams and coastal waters and affect both human and animal health through inhalation, skin, digestion, and eye irritation. Therefore, efficient and rapid removal of oil and solvent spills on the surface of water is imperative for protecting the aquatic environment, water resources, and subsequently human and animal health. Currently available oil and solvent spill techniques include physical diffusion, in-situ burning, bioremediation, and mechanical recovery [101, 102]. Among those technologies, cleaning the spilled oil and solvents by physical adsorbents is an attractive approach because of their ease of use and simplicity. Biopolymer-based adsorbents have been widely investigated for cleanup of oil and solvents due to their relative abundance, sustainability, low-cost, selectivity, fast adsorption kinetics, reusability and environmental benignity. Some recent developments in the use of biopolymer-based materials in oil and solvent spillage cleanup processes are discussed below.

Ao et al. reported the synthesis and separation of different oil/water mixtures using a cellulose hydrogel-coated mesh (CHM) [103]. They reported selective separation of a series of mixtures including hexane, cyclohexane, petroleum ether, liquid paraffin, pump oil and xylene with a high separation efficiency of over 98.9% and a high flux of 12,885 L/ m^2 h. Their material also showed high reusability, anti-salt properties, and stability after 60 successive cycles of separation with a separation efficiency of >98.2%. Magnetic hydrophobic polyvinyl alcohol (PVA)-cellulose nanofiber (CNF) aerogels (MHPCA) were prepared and used as effective oil absorbents by Xu and coworkers [104]. A series of oils including soybean oil, corn germ oil, pump oil, used pump oil, gasoline, motor oil, ethanol, and dimethylformamide

(DMF), were tested and MHPCA showed oil absorption capacity in the range of 59–136 g/g. MHPCA exhibited magnetically driven oil absorption and elasticity after 30 compression-release cycles, showing its reusability and durability. Cheng and coworkers investigated the use of pure cotton and cotton/cellulose aerogels derived from pure cotton and cellulose fiber from paper waste for the removal of oil and organic solvents [105]. The cotton/cellulose aerogels exhibited better performance over pure cotton aerogels because of the synergetic effects of two different cellulose fiber sources. The sorption capacity of the cotton/cellulose composite aerogel was 72.3 and 94.3 g/g for machine oil and dichloromethane, respectively [105].

Duan and coworkers prepared hydrophobic and oleophilic methyltrichlorosilane (MTCS) coated-chitin sponges and showed their ability to absorb a wide range of oils and nonpolar organic solvents including chloroform, gasoline oil, hexane, pump oil, silicon oil and toluene from the surface and bottom of the polluted water [106]. They reported that low viscosity organics such as chloroform, gasoline oil, hexane, and toluene could be absorbed to equilibrium within 8 min, whereas higher viscosity oils, such as pump oil and silicon oil could reach the absorption equilibrium within 13 min. MTCS-coated chitin sponges displayed recyclability with at least 10 times with approximately 93% adsorption capacity. Li et al. prepared porous chitosan-oxidized cellulose aerogel as an oil and organic solvent absorbent [107]. They reported the adsorption capacity in the range of 13.77–28.20 g/g for various oils and organic solvents including carbon tetrachloride, ethylene glycol, silicon oil, ethylene acetate, pump oil, crude oil, acetone, ethanol, diesel, and gasoline. Their material was stable over 50 absorption-desorption cycles following a pseudo-second-order kinetic model. The utilization of N,O-carboxymethyl chitosan for the adsorption of Marine Diesel, Diesel and Marine-2T oil for oil spill treatment has been reported by Doshi and coworkers [108].

5.4 Carbon dioxide adsorption

Carbon dioxide (CO_2) is the primary greenhouse gas (GHG) emitted through human activities and also considered as one of the major contributors to the greenhouse effect and global warming. CO_2 emissions, primarily from fossil fuel based power plants account for over 87% of the global annual emission of CO_2 [109]. According to the Intergovernmental Panel (IPCC) Fifth Assessment Report, the industrial revolution has significantly contributed to the raised atmospheric CO_2 levels from 280 to 400 parts per million (ppm) in the last 150 years [110]. It is estimated that by the end of year 2030, the global GHG emission could rise by 25–90% as compared to the GHG level in 2000, reaching CO_2 atmospheric levels in the range of 600–1550 ppm [111]. Some long-term effects of global warming include higher temperatures, changing rainfall patterns, rising sea levels, severe weather events ranging from flash floods, high intense hurricanes, freezing winters, severe droughts and heat waves. Therefore, many approaches including improved energy efficiency and energy conservation, increased utilization of low carbon fuels such as natural gas, hydrogen or nuclear power, usage of renewable energy such as solar, wind and hydropower and bioenergy, application of geoengineering processes such as afforestation and reforestation and use of carbon dioxide capture, storage and utilization (CCSU) techniques are extensively considered and adopted by various countries to mitigate CO_2 emissions [112, 113]. Among those, CCSU technologies have received a significant interest due to their ability to control in-situ CO_2 emissions from both industrial sources and natural sources. CCSU technologies mainly accompany separation of CO_2, transportation of captured and compressed CO_2, and underground storage of CO_2 for reuse. However, the success of these CCSU technologies relies on the CO_2 adsorption efficiency, manufacturing cost, and ease of handling and regeneration. Solid adsorbents include carbonaceous materials, covalent organic frameworks (COFs), porous organic frameworks (POPs) and metal–organic frameworks (MOFs), metal oxides (alkaline, alkaline earth and transition metals) and mesoporous silica-based materials and amines have been proposed for CCSU technologies [114–116]. Solid adsorbents derived from biopolymers are attracting interest for CCSU technologies because of their unique properties. Recently reported applications of biopolymer-based materials in capturing CO_2 gas are discussed in this section.

Biopolymers/biopolymers based composites	Pollutants	References
Cellulose	Acid blue 93 and methylene blue	[80]
	Malachite green	[81]
	Congo red	[82]
	Red 180, Blue CR 19, and Orange 142	[83]
	Ag(I) and Sn(II)	[95]
	Pb(II), Cd(II), Ni(II), and Zn(II)	[95, 96]
	Cu(II)	[95–97]
	Hexane, cyclohexane, petroleum ether	[104]
	Liquid paraffin, pump oil, and xylene	
	Machine oil and dichloromethane	[105]
	CO_2	[115, 117–119]
Chitin	Reactive yellow 2 and Reactive black 5	[84]
	Methylene blue	[85, 86]
	Indigo carmine	[87]
	Fe(III) and Cu(II)	[98]
	Chloroform, gasoline oil, hexane, toluene,	[106]
	silicon oil, pump oil	
	CO_2	[39]
Chitosan	Methylene blue	[88]
	Malachite green	[89]
	Acid red 73 and Pb(II)	[99]
	As(III) and As(IV)	[100]
	Carbon tetrachloride, ethylene glycol, silicon oil,	[107]
	Ethylene acetate, pump oil, crude oil, acetone, ethanol, and gasoline	[107]
	Diesel	[107, 108]
	Marine diesel	[108]
	CO_2	[120, 121]

Table 1.
Polysaccharide based biopolymers used in sorption of different pollutants.

Dassanayake et al. investigated the preparation of activated carbon monoliths derived from cellulose-based aerogel (aerocellulose) and its CO_2 adsorption properties at low and ambient temperatures [117]. The activated carbon prepared from aerocellulose exhibited a specific surface area of 753 m^2/g, a total pore volume of 0.72 cm^3/g, and a micropore volume of 0.27 cm^3/g. The physical CO_2 uptakes for these materials were 5.8 mmol/g of CO_2 at 0°C and 1 atm and 3.7 mmol/g of CO_2 at 25°C and 1.2 atm. They also reported the synthesis and CO_2 adsorption of a series of amidoxime (AO)-functionalized microcrystalline (MCC) and nanocrystalline cellulose (NCC)-mesoporous silica composites at ambient and elevated temperatures [115, 118]. They reported the CO_2 sorption capacities of MCC-AO composites in the range of 0.40–1.27 and 2.84–3.85 mmol/g at ambient (25°C, 1.2 atm) and elevated temperatures (120°C, 1.0 atm), respectively. Whereas, NCC-AO composites which showed highest CO_2 uptakes of 3.30 mmol/g at 25°C (1.2 atm) and 5.54 mmol/g at 120°C (1 atm), respectively. Both MCC-AO and NCC-AO composites displayed a good recyclability and stability after 10 successive adsorption/desorption cycles with negligible losses of the sorption capacity. Shehaqui et al. reported the direct CO_2 capture from air onto nanofibrillated cellulose (NFC)-polyethylenimine (PEI) foams [119]. They demonstrated the impact of both PEI and relative humidity (RH) on the CO_2 capture under atmospheric conditions with CO_2 concentration of $\sim$400 ppm. At 80% RH and PEI content of 44 wt%, a CO_2 uptake of 2.22 mmol/g was achieved with a stability over five repetitive cycles.

Dassanayake and coworkers also studied the CO_2 adsorption on activated carbon prepared by carbonization and KOH activation of chitin aerogels [39]. Their material showed CO_2 adsorption capacities of 5.02 mmol/g at 0°C and 3.44 mmol/g at 25° C under ambient pressure of 1 atm. Eftaiha et al. investigated the CO_2 adsorption by a chitin-acetate (CA)/dimethyl sulfoxide (DMO) binary system and reported the CO_2 uptake of 3.63 mmol/g at 3.95 atm and 0°C [120]. Fujiki and Yogo investigated CO_2 adsorption capacities of nitrogen-doped activated carbons prepared from chitosan at two different pressure conditions; namely 0.15 and 0.99 atm at 25°C [121]. They reported CO_2 adsorption capacities of 1.6 mmol/g at 0.15 atm and 4.9 mmol/g at 0.99 atm for their material. Alhwaige et al. reported the preparation of montmorillonite (MMT) reinforced bio-based chitosan-polybenzoxazine (CTS-PBZ) composite carbon aerogels and their ability to capture CO_2 [122]. MMT-CTS-PBZ composite carbon aerogels showed a maximum CO_2 adsorption of 5.72 mmol/g with multi-cyclic adsorption-desorption stability.

Table 1 summarizes the main applications of biopolymer-based materials for pollutants removal from water and sorption of CO_2.

6. Conclusion

Biopolymers and biopolymer-based materials have been widely investigated for various sorption applications due to their excellent adsorption capacities, easy functionalization, relative abundance, green, sustainable, biodegradable, low cost, and environmental benign properties. In this chapter, we presented the properties, processing, characterization and sorption applications of three most commonly studied natural polysaccharides, namely cellulose, chitin and chitosan. Recent developments of cellulose, chitin and chitosan-based materials in organic dye removal, heavy metals removal, oil and solvent spillage cleanup, and CO_2 adsorption have been extensively discussed.

Author details

Rohan S. Dassanayake, Sanjit Acharya and Noureddine Abidi*
Department of Plant and Soil Science, Fiber and Biopolymer Research Institute, Texas Tech University, Lubbock, TX, USA

*Address all correspondence to: noureddine.abidi@ttu.edu

References

[1] Pattanashetti NA, Heggannavar GB, Kariduraganavar MY. Smart biopolymers and their biomedical applications. Procedia Manufacturing. 2017;**12**:263-279

[2] Numata K. Poly(amino acid)s/polypeptides as potential functional and structural materials. Polymer Journal. 2015;**47**:537

[3] Klemm D. et al. Cellulose: Fascinating biopolymer and sustainable raw material. Angewandte Chemie International Edition. 2005;**44**(22):3358-3393

[4] Rinaudo M. Chitin and chitosan: Properties and applications. Progress in Polymer Science. 2006;**31**(7):603-632

[5] Lenz DM et al. Multiple reprocessing cycles of corn starch-based biocomposites reinforced with Curauá fiber. Journal of Polymers and the Environment. 2018

[6] Chen X et al. Effect of treatment methods on chitin structure and its transformation into nitrogen-containing chemicals. ChemPlusChem. 2015;**80**(10):1565-1572

[7] Hu Y, Abidi N. Distinct chiral nematic self-assembling behavior caused by different size-unified cellulose nanocrystals via a multistage separation. Langmuir. 2016;**32**(38):9863-9872

[8] Hu Y, Catchmark JM. In vitro biodegradability and mechanical properties of bioabsorbable bacterial cellulose incorporating cellulases. Acta Biomaterialia. 2011;7(7):2835-2845

[9] Hu Y, Catchmark JM, Vogler EA. Factors impacting the formation of sphere-like bacterial cellulose particles and their biocompatibility for human osteoblast growth. Biomacromolecules. 2013;**14**(10):3444-3452

[10] Hu Y et al. Engineering of porous bacterial cellulose toward human fibroblasts ingrowth for tissue engineering. Journal of Materials Research. 2014;**29**(22):2682-2693

[11] Hu Y et al. Preparation, characterization, and cationic functionalization of cellulose-based aerogels for wastewater clarification. Journal of Materials. 2016;**2016**:10

[12] Hu Y et al. Bioabsorbable cellulose composites prepared by an improved mineral-binding process for bone defect repair. Journal of Materials Chemistry B. 2016;**4**(7):1235-1246

[13] Wang S, Lu A, Zhang L. Recent advances in regenerated cellulose materials. Progress in Polymer Science. 2016;**53**:169-206

[14] Moon RJ et al. Cellulose nanomaterials review: Structure, properties and nanocomposites. Chemical Society Reviews. 2011;**40**(7):3941-3994

[15] Habibi Y, Lucia LA, Rojas OJ. Cellulose Nanocrystals: Chemistry, self-assembly, and applications. Chemical Reviews. 2010;**110**(6):3479-3500

[16] French AD. Glucose, not cellobiose, is the repeating unit of cellulose and why that is important. Cellulose. 2017;**24**(11):4605-4609

[17] George J, Sabapathi SN. Cellulose nanocrystals: Synthesis, functional properties, and applications. Nanotechnology, Science and Applications. 2015;**8**:45-54

[18] Gibson LJ. The hierarchical structure and mechanics of plant

materials. Journal of the Royal Society Interface. 2012;**9**(76):2749-2766

[19] Parthasarathi R et al. Insights into hydrogen bonding and stacking interactions in cellulose. The Journal of Physical Chemistry A. 2011;**115**(49): 14191-14202

[20] Peresin MS et al. Nanofiber composites of polyvinyl alcohol and cellulose nanocrystals: Manufacture and characterization. Biomacromolecules. 2010;**11**(3):674-681

[21] Khan F, Ahmad SR. Polysaccharides and their derivatives for versatile tissue engineering application. Macromolecular Bioscience. 2013;**13**(4):395-421

[22] Tanaka K, Yamamoto K, Kadokawa J-I. Facile nanofibrillation of chitin derivatives by gas bubbling and ultrasonic treatments in water. Carbohydrate Research. 2014;**398**:25-30

[23] Van de Velde K, Kiekens P. Structure analysis and degree of substitution of chitin, chitosan and dibutyrylchitin by FT-IR spectroscopy and solid state ^{13}C NMR. Carbohydrate Polymers. 2004;**58**(4):409-416

[24] Ravi Kumar MNV. A review of chitin and chitosan applications. Reactive and Functional Polymers. 2000;**46**(1):1-27

[25] Kurita K. Controlled functionalization of the polysaccharide chitin. Progress in Polymer Science. 2001;**26**(9):1921-1971

[26] Di Martino A, Sittinger M, Risbud MV. Chitosan: A versatile biopolymer for orthopaedic tissue-engineering. Biomaterials. 2005;**26**(30):5983-5990

[27] Acharya S et al. Chemical cationization of cotton fabric for improved dye uptake. Cellulose. 2014; **21**(6):4693-4706

[28] Pinkert A et al. Ionic liquids and their interaction with cellulose. Chemical Reviews. 2009;**109**(12): 6712-6728

[29] Wu J et al. Homogeneous acetylation of cellulose in a new ionic liquid. Biomacromolecules. 2004;**5**(2): 266-268

[30] Xiao P et al. Synthesis, characterization and properties of novel cellulose derivatives containing phosphorus: Cellulose diphenyl phosphate and its mixed esters. Cellulose. 2014;**21**(4):2369-2378

[31] Rana S et al. Regenerated cellulosic fibers and their implications on sustainability. In: Muthu SS, editor. Roadmap to Sustainable Textiles and Clothing: Eco-Friendly Raw Materials, Technologies, and Processing Methods. Singapore: Springer Singapore; 2014. pp. 239-276

[32] Heize T, El Seoud OA, Koschella A. Cellulose derivatives: Synthesis, structure, and properties. In: Springer Series on Polymer and Composite Materials. Springer International Publishing; 2018

[33] Duan B et al. Recent advances in chitin based materials constructed via physical methods. Progress in Polymer Science. 2018;**82**:1-33

[34] Budtova T, Navard P. Cellulose in NaOH–water based solvents: A review. Cellulose. 2016;**23**(1):5-55

[35] Potthast A et al. The cellulose solvent system N,N-dimethylacetamide/ lithium chloride revisited: The effect of water on physicochemical properties and chemical stability. Cellulose. 2002; **9**(1):41-53

[36] Sen S, Martin JD, Argyropoulos DS. Review of cellulose non-derivatizing solvent interactions with emphasis on activity in inorganic molten salt

hydrates. ACS Sustainable Chemistry & Engineering. 2013;**1**(8):858-870

[37] Vitz J et al. Extended dissolution studies of cellulose in imidazolium based ionic liquids. Green Chemistry. 2009;**11**(3):417-424

[38] Wang H, Gurau G, Rogers RD. Ionic liquid processing of cellulose. Chemical Society Reviews. 2012;**41**(4): 1519-1537

[39] Dassanayake RS et al. Activated carbon derived from chitin aerogels: Preparation and CO_2 adsorption. Cellulose. 2018;**25**(3):1911-1920

[40] Duan B et al. High strength films with gas-barrier fabricated from chitin solution dissolved at low temperature. Journal of Materials Chemistry A. 2013; **1**(5):1867-1874

[41] Acharya S. et al. Preparation and characterization of transparent cellulose films using an improved cellulose dissolution process. Journal of Applied Polymer Science. 2017;**134**(21)

[42] Acharya S, Hu Y, Abidi N. Mild condition dissolution of high molecular weight cotton cellulose in 1-butyl-3-methylimidazolium acetate/N,N-dimethylacetamide solvent system. Journal of Applied Polymer Science. 2018;**135**(9):45928

[43] Zhao S et al. Biopolymer aerogels: Chemistry, properties and applications. Angewandte Chemie International Edition. 2018;**57**:7580-7608

[44] Cai J et al. Cellulose aerogels from aqueous alkali hydroxide–urea solution. ChemSusChem. 2008;**1**(1–2):149-154

[45] Robitzer M, Renzo FD, Quignard F. Natural materials with high surface area. Physisorption methods for the characterization of the texture and surface of polysaccharide aerogels. Microporous and Mesoporous Materials. 2011;**140**(1):9-16

[46] Primo A, Quignard F. Chitosan as efficient porous support for dispersion of highly active gold nanoparticles: Design of hybrid catalyst for carbon-carbon bond formation. Chemical Communications. 2010;**46**(30): 5593-5595

[47] Lin N, Dufresne A. Nanocellulose in biomedicine: Current status and future prospect. European Polymer Journal. 2014;**59**:302-325

[48] Ifuku S. Chitin and chitosan nanofibers: Preparation and chemical modifications. Molecules. 2014;**19**(11): 18367-18380

[49] Dutta AK et al. Simple preparation of chitin Nanofibers from dry squid pen β-chitin powder by the star burst system. Journal of Chitin and Chitosan Science. 2013;**1**(3):186-191

[50] Kalia S et al. Nanofibrillated cellulose: Surface modification and potential applications. Colloid and Polymer Science. 2014;**292**(1):5-31

[51] Tzoumaki MV, Moschakis T, Biliaderis CG. Metastability of nematic gels made of aqueous chitin nanocrystal dispersions. Biomacromolecules. 2010; **11**(1):175-181

[52] Park S et al. Cellulose crystallinity index: Measurement techniques and their impact on interpreting cellulase performance. Biotechnology for Biofuels. 2010;**3**:10-10

[53] Liu S et al. Extraction and characterization of chitin from the beetle *Holotrichia parallela* Motschulsky. Molecules. 2012;**17**(4): 4604-4611

[54] Li L-H et al. Synthesis and characterization of chitosan/ZnO nanoparticle composite membranes. Carbohydrate Research. 2010;**345**(8): 994-998

[55] Abidi N, Cabrales L, Haigler CH. Changes in the cell wall and cellulose content of developing cotton fibers investigated by FTIR spectroscopy. Carbohydrate Polymers. 2014;**100**:9-16

[56] Dassanayake RS et al. One-pot synthesis of MnO_2–chitin hybrids for effective removal of methylene blue. International Journal of Biological Macromolecules. 2016;**93**:350-358

[57] Queiroz MF et al. Does the use of chitosan contribute to oxalate kidney stone formation? Marine Drugs. 2014; **13**:141-158

[58] Kumirska J et al. Application of spectroscopic methods for structural analysis of chitin and chitosan. Marine Drugs. 2010;**8**(5):1567-1636

[59] Chequer FMD et al. Textile dyes: Dyeing process and environmental impact. In: Günay M, editor. Eco-Friendly Textile Dyeing and Finishing. Rijeka: InTech; 2013. p. Ch. 06

[60] Yagub MT et al. Dye and its removal from aqueous solution by adsorption: A review. Advances in Colloid and Interface Science. 2014;**209**:172-184

[61] Zhou Y et al. Removal of methylene blue dyes from wastewater using cellulose-based superadsorbent hydrogels. Polymer Engineering & Science. 2011;**51**(12):2417-2424

[62] Li Y et al. Comparative study of methylene blue dye adsorption onto activated carbon, graphene oxide, and carbon nanotubes. Chemical Engineering Research and Design. 2013; **91**(2):361-368

[63] Zhu MX et al. Removal of an anionic dye by adsorption/precipitation processes using alkaline white mud. Journal of Hazardous Materials. 2007; **149**(3):735-741

[64] Alaoui A et al. Activity of Pt/MnO_2 electrode in the electrochemical degradation of methylene blue in aqueous solution. Separation and Purification Technology. 2015;**154**: 281-289

[65] Chan YJ et al. A review on anaerobic–aerobic treatment of industrial and municipal wastewater. Chemical Engineering Journal. 2009; **155**(1):1-18

[66] Işık M, Sponza DT. Biological treatment of acid dyeing wastewater using a sequential anaerobic/aerobic reactor system. Enzyme and Microbial Technology. 2006;**38**(7):887-892

[67] Patel H, Vashi RT. Removal of Congo red dye from its aqueous solution using natural coagulants. Journal of Saudi Chemical Society. 2012;**16**(2): 131-136

[68] Teh CY et al. Recent advancement of coagulation–flocculation and its application in wastewater treatment. Industrial & Engineering Chemistry Research. 2016;**55**(16):4363-4389

[69] Ciardelli G, Corsi L, Marcucci M. Membrane separation for wastewater reuse in the textile industry. Resources, Conservation and Recycling. 2001; **31**(2):189-197

[70] Huang J et al. Evaluation of micellar enhanced ultrafiltration for removing methylene blue and cadmium ion simultaneously with mixed surfactants. Separation and Purification Technology. 2014;**125**:83-89

[71] Zhang K et al. Comparison of catalytic activities for photocatalytic and sonocatalytic degradation of methylene blue in present of anatase TiO_2–CNT catalysts. Ultrasonics Sonochemistry. 2011;**18**(3):765-772

[72] Dariani RS et al. Photocatalytic reaction and degradation of methylene blue on TiO_2 nano-sized particles. Optik —International Journal for Light and Electron Optics. 2016;**127**(18):7143-7154

[73] Rosario-Ortiz FL, Wert EC, Snyder SA. Evaluation of UV/H_2O_2 treatment for the oxidation of pharmaceuticals in wastewater. Water Research. 2010; **44**(5):1440-1448

[74] Raghu S, Basha CA. Chemical or electrochemical techniques, followed by ion exchange, for recycle of textile dye wastewater. Journal of Hazardous Materials. 2007;**149**(2):324-330

[75] Zhuang L et al. A novel bioelectro-Fenton system for coupling anodic COD removal with cathodic dye degradation. Chemical Engineering Journal. 2010; **163**(1):160-163

[76] Lim H et al. Highly active heterogeneous Fenton catalyst using iron oxide nanoparticles immobilized in alumina coated mesoporous silica. Chemical Communications. 2006;(4): 463-465

[77] Hsueh CL et al. Degradation of azo dyes using low iron concentration of Fenton and Fenton-like system. Chemosphere. 2005;**58**(10):1409-1414

[78] Matheswaran M, Raju T. Destruction of methylene blue by mediated electrolysis using two-phase system. Process Safety and Environmental Protection. 2010;**88**(5): 350-355

[79] Dassanayake RS, Rajakaruna E, Abidi N. Preparation of aerochitin-TiO_2 composite for efficient photocatalytic degradation of methylene blue. Journal of Applied Polymer Science. 2018; **135**(8):45908

[80] Liu L et al. Adsorption removal of dyes from single and binary solutions using a cellulose-based bioadsorbent. ACS Sustainable Chemistry & Engineering. 2015;**3**(3):432-442

[81] Jiang F, Dinh DM, Hsieh Y-L. Adsorption and desorption of cationic malachite green dye on cellulose nanofibril aerogels. Carbohydrate Polymers. 2017;**173**:286-294

[82] Ruan C-Q, Strømme M, Lindh J. Preparation of porous 2,3-dialdehyde cellulose beads crosslinked with chitosan and their application in adsorption of Congo red dye. Carbohydrate Polymers. 2018;**181**: 200-207

[83] Maatar W, Boufi S. Microporous cationic nanofibrillar cellulose aerogel as promising adsorbent of acid dyes. Cellulose. 2017;**24**(2):1001-1015

[84] Akkaya G, Uzun İ, Güzel F. Kinetics of the adsorption of reactive dyes by chitin. Dyes and Pigments. 2007;**73**(2): 168-177

[85] Cao Y-L et al. Modification of chitin with high adsorption capacity for methylene blue removal. International Journal of Biological Macromolecules. 2018;**114**:392-399

[86] Xu R et al. Chitin/clay microspheres with hierarchical architecture for highly efficient removal of organic dyes. Carbohydrate Polymers. 2018;**188**: 143-150

[87] Gopi S et al. Chitin nanowhisker (ChNW)-functionalized electrospun PVDF membrane for enhanced removal of indigo carmine. Carbohydrate Polymers. 2017;**165**:115-122

[88] Xu B et al. Poly(2-acrylamido-2-methylpropane sulfonic acid) grafted magnetic chitosan microspheres: Preparation, characterization and dye adsorption. International Journal of Biological Macromolecules. 2018;**112**: 648-655

[89] Naseeruteen F et al. Adsorption of malachite green from aqueous solution by using novel chitosan ionic liquid beads. International Journal of Biological Macromolecules. 2018;**107**: 1270-1277

[90] Tchounwou PB et al. Heavy metals toxicity and the environment. EXS. 2012;**101**:133-164

[91] Carolin CF et al. Efficient techniques for the removal of toxic heavy metals from aquatic environment: A review. Journal of Environmental Chemical Engineering. 2017;**5**(3): 2782-2799

[92] Jamshaid A et al. Cellulose-based materials for the removal of heavy metals from wastewater—An overview. ChemBioEng Reviews. 2017;**4**(4): 240-256

[93] Anastopoulos I et al. Chitin adsorbents for toxic metals: A review. International Journal of Molecular Sciences. 2017;**18**(1):114

[94] Zhang L, Zeng Y, Cheng Z. Removal of heavy metal ions using chitosan and modified chitosan: A review. Journal of Molecular Liquids. 2016;**214**:175-191

[95] d'Halluin M et al. Chemically modified cellulose filter paper for heavy metal remediation in water. ACS Sustainable Chemistry & Engineering. 2017;**5**(2):1965-1973

[96] Fakhre NA, Ibrahim BM. The use of new chemically modified cellulose for heavy metal ion adsorption. Journal of Hazardous Materials. 2018;**343**:324-331

[97] Li Z et al. Converting untreated waste office paper and chitosan into aerogel adsorbent for the removal of heavy metal ions. Carbohydrate Polymers. 2018;**193**:221-227

[98] Duan Y et al. Lignin/chitin films and their adsorption characteristics for heavy metal ions. ACS Sustainable Chemistry & Engineering. 2018;**6**(5): 6965-6973

[99] Wu D et al. EDTA modified beta-cyclodextrin/chitosan for rapid removal of Pb(II) and acid red from aqueous solution. Journal of Colloid and Interface Science. 2018;**523**:56-64

[100] Kwok KCM et al. Adsorption/desorption of arsenite and arsenate on chitosan and nanochitosan. Environmental Science and Pollution Research. 2018;**25**(15):14734-14742

[101] Doshi B, Sillanpää M, Kalliola S. A review of bio-based materials for oil spill treatment. Water Research. 2018; **135**:262-277

[102] Tu L et al. *Calotropis gigantea* fiber derived carbon fiber enables fast and efficient absorption of oils and organic solvents. Separation and Purification Technology. 2018;**192**:30-35

[103] Ao C et al. Reusable, salt-tolerant and superhydrophilic cellulose hydrogel-coated mesh for efficient gravity-driven oil/water separation. Chemical Engineering Journal. 2018; **338**:271-277

[104] Xu Z et al. Preparation of magnetic hydrophobic polyvinyl alcohol (PVA)–cellulose nanofiber (CNF) aerogels as effective oil absorbents. Cellulose. 2018; **25**(2):1217-1227

[105] Cheng H et al. Cotton aerogels and cotton-cellulose aerogels from environmental waste for oil spillage cleanup. Materials & Design. 2017;**130**: 452-458

[106] Duan B et al. Hydrophobic modification on surface of chitin sponges for highly effective separation of oil. ACS Applied Materials & Interfaces. 2014;**6**(22):19933-19942

[107] Li Z et al. Excellent reusable chitosan/cellulose aerogel as an oil and organic solvent absorbent. Carbohydrate Polymers. 2018;**191**: 183-190

[108] Doshi B et al. Effectiveness of N,O-carboxymethyl chitosan on

destabilization of marine diesel, diesel and marine-2T oil for oil spill treatment. Carbohydrate Polymers. 2017;**167**: 326-336

[109] Le Quéré C et al. The global carbon budget 1959–2011. Earth System Science Data. 2012;**5**(2):1107-1157

[110] McHale G, Shirtcliffe NJ, Newton MI. Contact-angle hysteresis on super-hydrophobic surfaces. Langmuir. 2004; **20**(23):10146-10149

[111] Cassie ABD, Baxter S. Wettability of porous surfaces. Transactions of the Faraday Society. 1944;**40**:546-551

[112] Wang Q et al. CO_2 capture by solid adsorbents and their applications: Current status and new trends. Energy & Environmental Science. 2011;**4**(1): 42-55

[113] Leung DYC, Caramanna G, Maroto-Valer MM. An overview of current status of carbon dioxide capture and storage technologies. Renewable and Sustainable Energy Reviews. 2014; **39**:426-443

[114] D'Alessandro DM, Smit B, Long JR. Carbon dioxide capture: Prospects for new materials. Angewandte Chemie (International Ed. in English). 2010; **49**(35):6058-6082

[115] Dassanayake RS et al. Amidoxime-functionalized nanocrystalline cellulose-mesoporous silica composites for carbon dioxide sorption at ambient and elevated temperatures. Journal of Materials Chemistry A. 2017;**5**(16): 7462-7473

[116] Gregor S, Ganin AY, Yiu HHP. Sustainable CO_2 adsorbents prepared by coating chitosan onto mesoporous silicas for large-scale carbon capture technology. Energy Technology. 2015; **3**(**3**):249-258

[117] Dassanayake RS et al. Preparation and adsorption properties of aerocellulose-derived activated carbon monoliths. Cellulose. 2016;**23**(2): 1363-1374

[118] Gunathilake C et al. Amidoxime-functionalized microcrystalline cellulose-mesoporous silica composites for carbon dioxide sorption at elevated temperatures. Journal of Materials Chemistry A. 2016;**4**(13):4808-4819

[119] Sehaqui H et al. Fast and reversible direct CO_2 capture from air onto all-polymer nanofibrillated cellulose—polyethylenimine foams. Environmental Science & Technology. 2015;**49**(5): 3167-3174

[120] Eftaiha AF et al. Chitin-acetate/DMSO as a supramolecular green CO_2-phile. RSC Advances. 2016;**6**(26): 22090-22093

[121] Fujiki J, Yogo K. The increased CO_2 adsorption performance of chitosan-derived activated carbons with nitrogen-doping. Chemical Communications. 2016;**52**(1):186-189

[122] Alhwaige AA, Ishida H, Qutubuddin S. Carbon aerogels with excellent CO_2 adsorption capacity synthesized from clay-reinforced biobased chitosan-polybenzoxazine nanocomposites. ACS Sustainable Chemistry & Engineering. 2016;**4**(3): 1286-1295

Section 2

Sorption Processes for the Removal of Various Pollutants

Chapter 2

Sorption of Phosphorus from Fertilizer Mixture

Augustine Muwamba, Kelly T. Morgan and Peter Nkedi-Kizza

Abstract

Studying phosphorus (P) sorption behavior is among the prerequisites for P management in the crop fields. The work presented in this chapter described P sorption data when fertilizer mixture (NH_4NO_3, KH_2PO_4, and KCl) was used to characterize sorption on soil. In addition to using fertilizer mixture, sorption experiments were also conducted using KH_2PO_4 prepared in 0.01 M KCl, 0.005 M $CaCl_2$, and deionized water. The 24-h batch sorption experiments were conducted using a sandy soil to solution ratio of 1:2, and the equilibrium solution and sorbed data were described using Freundlich isotherm. Sorption kinetics experiments were conducted using times, 4, 8, 12, and 24 h. The Freundlich isotherm constant and sorbed P kinetics data for 0.005 M $CaCl_2$ were significantly greater ($p < 0.05$) than for 0.01 M KCl and/or fertilizer mixture. The Freundlich isotherm constant and sorbed P kinetics data for deionized water were significantly lower ($p < 0.05$) than for 0.01 M KCl and/or fertilizer mixture. There was no significant difference in Freundlich isotherm constant and sorbed P kinetics data for 0.01 M KCl and fertilizer mixture. The sorption data showed the importance of using the fertilizer mix applied to the field when conducting sorption experiments.

Keywords: fertilizer mixture, isotherm, sorption coefficient, sorption kinetics

1. Introduction

Phosphorus (P) is applied with different nutrients to crop fields. Examples of field crops that need fertilizer mixture are shown in **Table 1**. Varying nutrients combinations can significantly affect the interactions of P with soil due to varying ionic strength and pH [1–7]. For example ionic strength was positively correlated to P sorption [2]. The specific affinity and the valence of the cation on the soil exchange site were also associated to P sorption capacity [7]. Supporting electrolytes are used for conducting P sorption experiments assuming representation of the true chemistry of the field solutions without necessarily considering the varying fertilizer mix applied to the soil. **Table 2** shows examples of supporting electrolytes that were used to characterize P sorption in the past studies. In this chapter, it was hypothesized that P sorption isotherm constants and kinetics data for fertilizer mix were significantly different from supporting electrolytes commonly used for conducting P sorption.

Sorption isotherms are used to describe relationships between sorbed and solution P in a given sorption experiment at constant temperature and act as indicators

Field crop	Fertilizer mixture distribution
Sugarcane	200 kg N, 50 kg P_2O_5, and 200 kg K_2O per acre
Canola spring type	160 lb N, 30 lb P_2O_5, and 40 lb K_2O per acre
Canola winter type	175 lb N, 30 lb P_2O_5, and 40 lb K_2O per acre
Corn (for grain) dryland	120 lb N, 20 lb P_2O_5, and 20 lb K_2O per acre
Corn (for grain) irrigated	180 lb N, 70 lb P_2O_5, and 70 lb K_2O per acre
Cotton (1500 lb yield goal)	105 lb N, 140 lb P_2O_5, and 80 lb K_2O per acre
Grain sorghum	80 lb N, 80 lb P_2O_5, and 80 lb K_2O per acre
Peanuts	0 lb N, 80 lb P_2O_5, and 80 lb K_2O per acre
Small grain-barley	100 lb N, 80 lb P_2O_5, and 80 lb K_2O per acre
Small grain-oats	105 lb N, 80 lb P_2O_5, and 80 lb K_2O per acre
Small grain-cover crop	60 lb N, 80 lb P_2O_5, and 80 lb K_2O per acre
Small grain-wheat	120 lb N, 80 lb P_2O_5, and 80 lb K_2O per acre
Small grain silage	160 lb N, 100 lb P_2O_5, and 160 lb K_2O per acre
Sorghum silage	150 lb N, 80 lb P_2O_5, and 160 lb K_2O per acre
Soybeans	0 lb N, 70 lb P_2O_5, and 100 lb K_2O per acre
Sunflower	80 lb N, 80 lb P_2O_5, and 80 lb K_2O per acre
Sweet sorghum	80 lb N, 80 lb P_2O_5, and 80 lb K_2O per acre
Tobacco	50 lb N, 100 lb P_2O_5, and 180 lb K_2O per acre
Kenaf	175 lb N, 100 lb P_2O_5, and 100 lb K_2O per acre
Truffles	50 lb N, 80 lb P_2O_5, and 80 lb K_2O per acre

Table 1.
Field crop and fertilizer mixture distributions.

of field P retention potential [24–27]. Sorption coefficient is among the coefficients described in the isotherms that is used to model P movement in the field [28]. Phosphorus sorption capacity has also been used as an important management tool in many crop fields [29]. Therefore, there is a need to identify the appropriate chemistry of the field solutions before conducting P sorption experiments and modeling P movement in the crop fields. Sorption kinetics data trends were reported to provide clues on the mechanisms of sorption reactions [30]; appropriate solution chemistry should also be carefully chosen for the sorption kinetics experiments. It was also hypothesized that the isotherms that describe sorption data from fertilizer mixture are different from isotherms that describe data from typical laboratory supporting electrolytes.

The importance of laboratory P sorption and kinetics data in modeling and understanding of the P dynamics in crop fields has been documented [31, 32]. The P sorption characteristics help to properly calibrate theoretical models that aim at mimicking field processes [31, 32]. Accurate laboratory sorption data collected using true field solution chemistry will therefore improve models as predictive tools for P movement. The objective of the study was to determine the differences in P sorption behavior for P in fertilizer mixture (N, P, and K) prepared in deionized water and in P fertilizer (KH_2PO_4) prepared in 0.01 MKCl, 0.005 M $CaCl_2$, and deionized water.

Electrolyte	Isotherm	Soil	Reference
0.1 M $CaCl_2$	ND	- Typic Argiudolls - Typic Hapludolls - Entic Haplustolls - Abruptic Argiudolls - Petrocalcic Paleudolls	[8]
0.01 M $CaCl_2$	- Langmuir - Freundlich - Temkin	- Vertisol - Rhizospheric soil - Haploboroll - Hapludoll - Eutrochrept - Haplaquept - Brunic Arenosols (dystric) - Haplic Regosol (dystric) - Sandy mixed Humic Dystrochrept - Very fine, mixed, semiactive, Oxyaquic Haplocryoll - Fine, illitic, frigid Typic Haplquept - Coarse-loamy, mixed, mesic Oxyaquic, Eutrochrept	[1, 9–14]
0.001 M $CaCl_2$	- Freundlich	- Haplustalf - Orthent - Tropaquept	[15]
0.01 M KCl	- Langmuir - Linear	- Sandy, siliceous, hyperthermic Aeric Alaquods - Alaquods and Alorthods - Loamy, siliceous, subactive, thermic Arenic Paleudults - Fine-loamy, siliceous, subactive, thermic Aquic Paleudults - Fine, mixed, semiactive, thermic Typic Umbraquults - Fine-loamy, mixed, semiactive, acid, thermic Histic Humaquepts	[16–19]
0.02 M KCl	- Langmuir	- Loamy, siliceous, hyperthermic Arenic Glossaqualf	[20]
0.05 M KCl	- Langmuir - Freundlich - Linear	- Quartzipsamments - Paleudults - Loamy-skeletal, carbonatic hyperthermic Lithic Udorthents - Loamy, carbonatic, hyperthermic, shallow Typic Fluvaquents - Loamy, skeletal, carbonatic, hyper thermic, Lithic Udorthents	[4, 21, 22]
0.1 M $NaNO_3$	- Langmuir	- Sandy, siliceous, hyperthermic Ultic Alaquod	[23]
0.1 M NaCl	ND	- Aquic or Oxyaquic Haplocryods	[3]
Deionized water	ND	- Aquic or Oxyaquic Haplocryods	[3]

ND, not determined.

Table 2.
Supporting electrolytes, soils, and sorption isotherms for the literature studies.

2. Sorption experiments and trends in sorption data

2.1 Determination of soil properties

The soil samples used for sorption experiments were air dried, passed through 2-mm sieve and first analyzed for pH, total carbon, oxalate extractable iron, oxalate extractable aluminum, and exchangeable calcium. Particle size distribution of the soil samples was also determined. A soil to water solution ratio of 1:2 was prepared, and the soil pH was measured using a standardized pH meter (model: AR15; manufacturer: Fisher Scientific) [33]. The combustion method with the element analyzer (Carbo-Erba NA 2500 instrument (Model: NA 2500; manufacturer: CE instruments, Italy) was used to measure total carbon. The inductively coupled plasma (ICP) (model: Optima 700 DV; manufacturer: Perkin Elmer) was used to analyze for oxalate iron and aluminum after extraction with oxalate solution [34]. Exchangeable calcium was also analyzed using ICP after extraction with 0.2 M NH_4Cl [35]. Particle size distribution was determined by hydrometer method [35] .

2.2 Determination of sorption isotherms

An example of P sorption experiment that involved using KH_2PO_4 fertilizer prepared in 0.01 M KCl, 0.005 M $CaCl_2$, and deionized water and fertilizer mixture (NH_4NO_3, KH_2PO_4, and KCl) prepared in deionized water was used for this study. The fertilizer rates, 50 kg P_2O_5 ha^{-1}, 200 kg N ha^{-1}, and 200 kg K_2O ha^{-1}, applied to sugarcane fields were used to prepare the fertilizer mixture. Potassium chloride and calcium chloride were used because they are commonly used to conduct P sorption experiments with the assumption that the solutions' ionic strength and pH are close to those of the crop fields. The concentration, 0.005 M for $CaCl_2$ and 0.01 M for KCl were used to attain the equivalences of Ca^{2+} and K^+. Deionized water was used because irrigation water is used to provide the necessary plant moisture.

The two sandy soils, Margate (sandy, siliceous, hyperthermic Mollic Psammaquents) and Immokalee (sandy, siliceous, hyperthermic Arenic Alaquods) used for the experiment contribute the most to the greatest percentage of soils used for sugarcane production in Southwestern Florida. Five soil samples of each of the soil horizons, A and Bh for Immokalee soil and A and Bw for Margate soil, were used to represent the varying soil properties (e.g., total carbon, iron, and aluminum). The soil samples were sampled from two sugarcane fields each of 12 ha located in Hendry County, southwestern Florida (26.75° N, 80.93° W).

The initial P concentrations (C0) used for the experiment ranged from 8 to 60 mg L^{-1}. The soil to solution ratio of 1:2 (10 g of soil and 20 mL of solution) was used and equilibrium solution concentration was analyzed after 24 h of shaking. Blanks where soil was shaken with only 0.01 M KCl, 0.005 M $CaCl_2$, and deionized water were also included in the experiment, and the blank equilibrium concentrations were subtracted from the treatment sample equilibrium concentrations. The experiments were conducted at room temperature (25°C). Before analyzing the solution concentrations, soil solutions were centrifuged at 5000 rpm for about 20 min and filtered using 42 Whatman filter. The spectrophotometer (HACH DR/4000U) was used to analyze solution P at a detection wavelength of 880 nm.

Sorbed P (S) was equal to V/M (C0 − Ce) where V, M, C0, and Ce are volume of solution, mass of soil, initial solution P concentration, and equilibrium P concentration, respectively. Sorption data for all the supporting electrolytes were fitted to Freundlich isotherm [sorbed (S) versus equilibrium solution (C) concentration]. The Freundlich sorption isotherm is represented by Eq. (1).

$$S = K_f C^N \tag{1}$$

where S is the amount of P sorbed (mg kg^{-1}), C is the solution P concentration (mg L^{-1}), K_f is the Freundlich sorption coefficient (L^N, kg^{-1} mg^{1-N}), and N is an empirical constant. The coefficients for fertilizer mixture were compared to coefficients for 0.01 M KCl, 0.005 M $CaCl_2$, and deionized water. The paired t-test was used to identify significant differences in Freundlich sorption coefficients.

2.3 Sorption kinetics experiments

The sorption of P has been assumed as a kinetic process [15, 28]. Three initial concentrations (C0), 19, 29, and 38 mg L^{-1} were used for conducting sorption kinetics experiment. The A horizon of Immokalee soil was used for sorption kinetics experiment, and the soil to solution ratio of 1:2 (10 g of soil and 20 mL of solution) was used. Solution concentrations were analyzed after 4, 8, 12, and 24 h. The paired t-test was used to identify significant differences in relative concentrations (C/C0) and sorbed concentrations between fertilizer mixture and supporting electrolytes. R-software was used for statistical analyses. Graphs of relative concentrations (C/C0) and sorbed concentrations (S) versus time were plotted to show the data trends over a 24 h period.

2.4 Selected soil properties

The average percent sand, pH (1:2 soil:water volume), total carbon, oxalate iron, oxalate aluminum, and exchangeable Ca for A horizon of Immokalee soil were 97.0%, 6.8, 15.2 g kg^{-1}, 234.0 mg kg^{-1}, 280.4 mg kg^{-1}, and 3.6 cmolc kg^{-1}, respectively. The average percent sand, pH (1:2 soil:water volume), total carbon, oxalate iron, oxalate aluminum, and exchangeable Ca for A horizon of Margate soil were 97.5.0%, 8.3, 11.2 g kg^{-1}, 661.1 mg kg^{-1}, 307.3 mg kg^{-1}, and 6.6 cmolc kg^{-1}, respectively. The average percent sand, pH (1:2 soil:water volume), total carbon, oxalate iron, oxalate aluminum, and exchangeable Ca for Bh were 87.5%, 6.8, 39.7 g kg^{-1}, 114.4 mg kg^{-1}, 305.0 mg kg^{-1}, and 5.4 cmolc kg^{-1}, respectively. The average percent sand, pH (1:2 soil:water volume), total carbon, oxalate iron, oxalate aluminum, and exchangeable Ca for Bw were 97.2%, 8.4, 3.9 g kg^{-1}, 149.0 mg kg^{-1}, 89.0 mg kg^{-1}, and 2.0 cmolc kg^{-1}, respectively.

2.5 Changes of sorption isotherm coefficients with supporting electrolytes

Although all sorption data fitted Freundlich isotherms with R^2 values greater than 0.9, the Freundlich coefficients varied with the type of supporting electrolytes (**Table 3**). For both 0.01 M KCl and fertilizer mixture, the Freundlich isotherm constant was significantly lower ($p < 0.05$) than for 0.005 M $CaCl_2$ and significantly greater ($p < 0.05$) than for deionized water (**Table 4**). Although the same equivalence was used for K^+ and Ca^{2+}, sorption was greater for 0.005 M $CaCl_2$ than

Soil	Horizon	0.01 M KCl	0.005 M $CaCl_2$	Deionized water	Fertilizer mixture
Immokalee soil	A	$S = 4.8\ C^{0.5}$	$S = 13.5\ C^{0.4}$	$S = 2.7C^{0.8}$	$S = 4.6C^{0.5}$
Margate soil	A	$S = 7.2\ C^{0.6}$	$S = 24.6\ C^{0.3}$	$S = 5.4C^{0.5}$	$S = 7.3\ C^{0.6}$
Immokalee soil	Bh	$S = 19.1C^{0.6}$	$S = 79.0C^{0.2}$	$S = 13.1C^{0.8}$	$S = 21.1C^{0.6}$
Margate soil	Bw	$S = 10.3C^{0.3}$	$S = 28.0\ C^{0.2}$	$S = 6.7C^{0.4}$	$S = 9.5C^{0.3}$

Table 3.
Average sorption isotherms of five replicates showing variabilities in Freundlich sorption coefficients.

Comparisons	A-Immokalee	A-Margate	Bh	Bw
Fertilizer mixture versus deionized water	S	S	S	S
Fertilizer mixture versus 0.005 M $CaCl_2$	S	S	S	S
Fertilizer mixture versus 0.01 M KCl	NS	NS	NS	NS
NS, no significant difference (α = 0.05); S, significantly greater or lower (α = 0.05).				

Table 4.
Comparisons of Freundlich sorption coefficients and sorbed phosphorus concentrations.

for 0.01 M KCl and fertilizer mixture likely due to the influence of charge (+2) on Ca^{2+} that reduce the electrostatic repulsion effect between phosphate and the soil surface. A similar trend ($Ca^{+2} > K^{+}$) for P sorption was identified in other sorption studies [36]. Phosphorus sorption was also reported to increase with increase in background electrolyte concentration [37]. The sorption characteristics of P (KH_2PO_4) prepared in deionized water was lower than for 0.01 M KCl and fertilizer mixture probably due to significantly lower K^{+} concentration that contributes less to ionic strength than in the latter two. The greater Freundlich coefficients for Bh were greater than A horizon because of the greater total carbon, oxalate iron, and oxalate aluminum that enhance greater P sorption. Soils with greater free aluminum and iron were associated with greater P sorption by different researchers [38–40].

2.6 Trends in sorption kinetics data for different supporting electrolytes

Figure 1 shows the sorbed P concentrations as a function of initial concentrations for a 24 h time step with no significant difference in sorbed concentrations for 0.01 M KCl and fertilizer mixture. **Figure 2** shows the trends in relative solution concentrations (C/C0) and sorbed concentrations (P). While the relative solution concentrations decreased over time, sorbed concentrations increased over time. For both 0.01 M KCl and fertilizer mixture, sorbed P kinetics data were significantly lower ($p < 0.05$) than for 0.005 M $CaCl_2$, and significantly greater ($p < 0.05$) than for deionized water (**Table 4**). Sorption was fast for the first hours due to the presence of high P affinity sorption sites on the exchange sites and gradual for the following hours (**Figure 2**). A fast P sorption first phase followed by a steady phase was also documented in other studies [2, 15, 41].

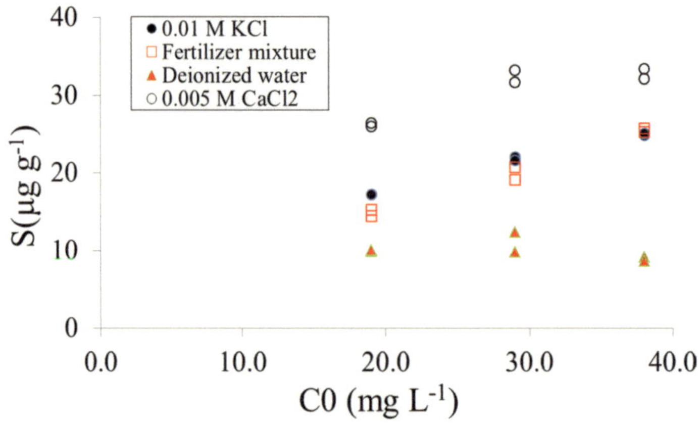

Figure 1.
Sorbed P (S) concentration as a function of initial concentrations (Co) for A horizon of Immokalee soil.

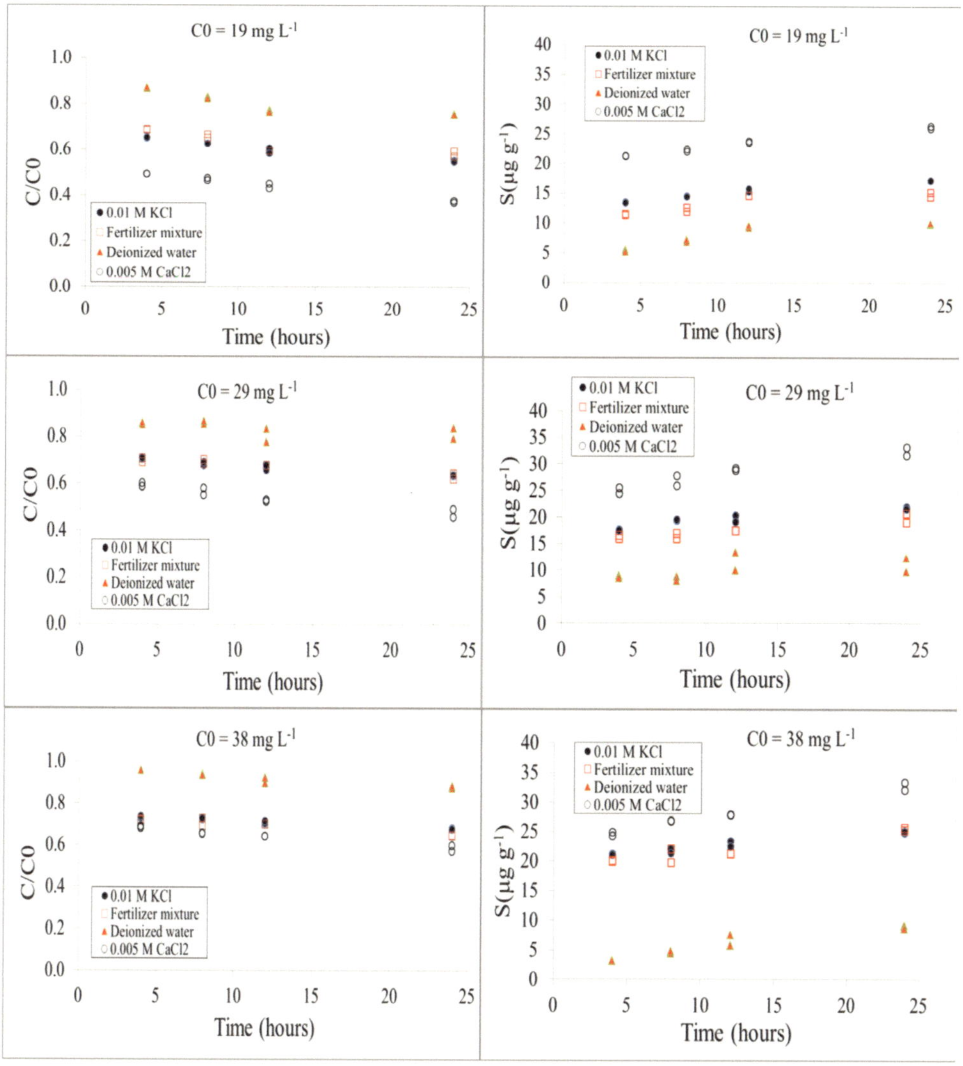

Figure 2.
Relative solution (C/Co) and sorbed (S) concentration as a function of time for A horizon of Immokalee soil.

3. Summary/conclusions

The results presented in this chapter suggest that if another nutrient is applied with P in the field, the P sorption behavior should be studied with the applied fertilizer mix, and P prepared in recommended supporting electrolyte as well. The sorption characterization with the two scenarios will help in identifying the appropriate sorption characteristics (sorption isotherm coefficients and kinetics constants) used for predicting P movement and P management options.

Acknowledgements

The authors are grateful to Florida Department of Environmental Regulation for providing the necessary funds.

Author details

Augustine Muwamba*, Kelly T. Morgan and Peter Nkedi-Kizza
University of Florida, Gainesville, Florida, USA

*Address all correspondence to: mati@ufl.edu

References

[1] Ahmed MF, Kennedy IR, Choudhury ATMA, Kecske ML, Deaker R. Phosphorus adsorption in some Australian soils and influence of bacteria on the desorption of phosphorus. Communications in Soil Science and Plant Analysis. 2008;**39**:1269-1294

[2] Wang S, Jin X, Bu Q, Zhou X, Wu F. Effect of particle size, organic matter and ionic strength of phosphate sorption in different trophic lake sediments. Journal of Hazardous Materials. 2006:95-105

[3] Giesler R, Anderson T, Lovgren L, Persson P. Phosphate sorption in aluminum- and iron-rich humus soils. Soil Science Society of America Journal. 2005;**69**(1):77-86

[4] Violante A, Pigna M. Competitive sorption of arsenate and phosphate on different clay minerals and soils. Soil Science Society of America Journal. 2002;**66**(6):1788-1796

[5] Curtin D, Syers JK, Bolan NS. Phosphate sorption by soil in relation to exchangeable cation composition and pH. Australian Journal of Soil Research. 1992;**31**:137-149

[6] Froelich PN. Kinetic control of dissolved phosphate in natural rivers and estuaries: A primer on the phosphate buffer mechanism. Limnology and Oceanography. 1988;**33**(4 pt 2):649-668

[7] Barrow NJ, Bowden JW, Posner AM, Quirk JP. Describing the effects of electrolyte on adsorption of phosphate by a variable charge surface. Australian Journal of Soil Research. 1980;**18**:395-404

[8] Rubio G, Cabello MJ, Boem FHG, Munaro E. Estimating available soil phosphorus increases after phosphorus additions in Mollisols. Soil Science Society of America Journal. 2008;**72**:1721-1727

[9] Börling K, Otabbong E, Barberis E. Phosphorus sorption in relation to soil properties in some cultivated Swedish soils. Nutrient Cycling in Agroecosystems. 2001;**59**:39-46

[10] Debicka M, Kocowicz A, Weber J, Jamroz E. Organic matter effects on phosphorus sorption in sandy soils. Archives of Agronomy and Soil Science. 2015. DOI: 10.1080/03650340.2015.1083981

[11] Daly K, Jeffrey D, Tunney H. The effect of soil type on phosphorus sorption capacity and desorption dynamics in Irish grassland soils. Soil Use and Management. 2001;**17**:12-20

[12] Djodjic F, Borling K, Bergstrom L. Phosphorus leaching in relation to soil type and soil phosphorus content. Journal of Environmental Quality. 2004;**33**(2):678-684

[13] Maguire RO, Foy RH, Bailey JS, Sims JT. Estimation of the phosphorus sorption capacity of acidic soils in Ireland. European Journal of Soil Science. 2001;**52**:479-487

[14] McDowell R, Condron L. Influence of soil constituents on soil phosphorus sorption and desorption. Communications in Soil Science and Plant Analysis. 2001;**32**(15, 16):2531-2547

[15] Agbenin JO, Tiessen H. Phosphorus sorption at field capacity and soil ionic strength: Kinetics and transformation. Soil Science Society of America Journal. 1995;**59**:998-1005

[16] Reddy KR, O'Connor GA, Gale PM. Phosphorus sorption capacities of wetlands soils and stream

sediments impacted by diary effluent. Journal of Environmental Quality. 1998;**27**:438-447

[17] Villapando RR, Graetz DA. Phosphorus sorption and desorption properties of the spodic horizon from selected Florida Spodosols. Soil Science Society of America Journal. 2001;**65**:331-339

[18] Kang J, Hesterberg D, Osmond DL. Soil organic matter effects on phosphorus sorption: A path analysis. Soil Science Society of America Journal. 2009;**73**(2):360-366

[19] Ige DV, Akinremi OO, Flatten DN. Direct and indirect effects of soil properties on phosphorus retention capacity. Soil Science Society of America Journal. 2007;**71**(1):95-100

[20] He ZL, Alva AK, Li YC, Calvert DV, Banks DJ. Sorption-desorption and solution concentration of phosphorus in a fertilized sandy soil. Journal of Environmental Quality. 1999;**28**:1804-1810

[21] Zhou M, Li Y. Phosphorus-sorption characteristics of calcareous soils and limestone from the southern Everglades and adjacent farmlands. Soil Science Society of America Journal. 2001;**65**:1404-1412

[22] Rhue RD, Harris WG, Nair VD. A retardation-based model for phosphorus transport in sandy soil. Soil Science. 2006;**171**:293-304

[23] Bhatti JS, Comerford NB, Johnston CT. Influence of oxalate and soil organic matter on sorption and desorption of phosphate onto a spodic horizon. Soil Science Society of America Journal. 1998;**62**:1089-1095

[24] Chen Z, Xing B, McGill WB. A unified sorption variable for environmental applications of the Freundlich equation. Journal of Environmental Quality. 1999;**28**(5):1422-1128

[25] Idris OAA, Ahmed HS. Phosphorus sorption capacity as a guide for phosphorus availability of selected Sudanese soil series. African Crop Science Journal. 2012:59-65

[26] Travis SC, Etnier EL. A survey of sorption relationships for reactive solutes in soil. Journal of Environmental Quality. 1981;**10**(1):8-17

[27] Richardson CJ. Mechanisms controlling phosphorus retention capacity in freshwater wetlands. American Association for the Advancement of Science. 1985;**228**(4706):1424-1427

[28] Šimůnek J, van Genuchten MT. Modeling nonequilibrium flow and transport processes using HYDRUS. Vadose Zone Journal. 2008;**7**(2):782-797

[29] Beauchemin S, Simard RR. Soil phosphorus saturation degree: Review of some indices and their suitability for P management in Québec, Canada. Canadian Journaal of Soil Science. 1999:615-628

[30] Wang Y, Shen Z, Niu J, Liu R. Adsorption of phosphorus on sediments from the three gorgeous reservoir (China) and the relation with sediment compositions. Journal of Hazardous Materials. 2009;**162**:92-98

[31] Simunek J, Sejna J, van Genuchten MT. The HYDRUS-2D Software Package for Simulating the Two and One Dimensional Movement of Water, Heat, and Multiple Solutes in Variably Saturated Media. USDA, Riverside, CA: US Salinity Laboratory Agricultural Research Service; 1999

[32] Hassan G, Reneau RB, Hagedorn C, Jantrania AR. Modeling effluent distribution and nitrate transport

through an on-site wastewater system. Journal of Environmental Quality. 2008;**37**:1937-1948

[33] Sparks DL. Soil Science Society of America Series (SSSA) 5: Methods of Soil Analysis. 1996

[34] McKeague JA, Day JH. Dithionite and oxalate extractable Fe and Al as aids in differentiating various classes of soils. Canadian Journal of Soil Science. 1966;**46**:13-22

[35] Soil Survey Staff. Keys to Soil Taxonomy. Washington, DC: U.S. Gov. Print. Office; 1996

[36] Pardo MT, Guadalix ME, Garcia-Gonzalez MT. Effect of pH and background electrolyte on P sorption by variable charge soils. Geoderma. 1992;**54**:275-284

[37] Antoniadis V, Koliniati R, Efstratiou E, Golia E, Petropoulos S. Effect of soils with varying degree of weathering and pH values on phosphorus sorption. Catena. 2016;**139**:214-219

[38] Gerard F. Clay minerals, iron/aluminum oxides, and their contribution to phosphate sorption in soils—A myth revisted. Geoderma. 2016;**262**:213-226

[39] Kisinyo PO, Othieno CO, Gudu SO, Okalebo JR, Opala PA, Maghanga JK, Ng'etich WK, Agalo JJ, Opile RW, Kisinyo JA, Ogola BO. Phosphorus sorption and lime requirements of maize growing acid soils of Kenya. Sustainable Agriculture Research. 2013;**2**(2):116-123

[40] Wolde Z, Haile W. Phosphorus soprtion isotherms and external phosphorus requirements of some soils of Southern Ethiopia. African Crop Science Journal. 2015;**23**(2):89-99

[41] Appan A, Wang H. Sorption isotherms and kinetics of sedimentphosphorus in a tropical reservoir. Journal of Environmental Engineering. 2000:993-998

Chapter 3

Calcium Uptake on Kaolinite and Gibbsite: Effects of Sulfate, pH, and Salt Concentration with Additional Insight from Second Harmonic Generation on Temperature Dependencies with Sapphire-Basal Planes and the Potential Relevance to Ice Nucleation

Ahmed Abdelmonem, Yujun Wang, Johannes Lützenkirchen and Marcelo Eduardo Alves

Abstract

Although previous studies have shown that sulfate can either increase cation leaching or enhance cation adsorption in soil, little is known about the factors behind these phenomena. To learn more about them, calcium adsorption experiments were carried out with kaolinite and gibbsite at initial pH values 4 and 6 and in the presence of 1 or 20 $mmol_c\ L^{-1}$ of either nitrate or sulfate. The results indicated that limited sulfate-calcium coadsorption occurred on gibbsite when it was in contact with the dilute solution of $CaSO_4.2H_2O$ at pH ~ 7. Regarding mineral and pH values, calcium adsorption from the concentrated solutions decreased with sulfate possibly because of the presence of ~31% of the $CaSO_4^0$ ion pair in the concentrated $CaSO_4.2H_2O$ solutions and the low free calcium activity therein. Calcium adsorption on kaolinite and gibbsite from all concentrated solutions was reduced when the initial pH changed from 4 to 6 suggesting a negative salt effect on that process. In addition to indicating negligible participation of gibbsite in calcium adsorption, our findings also suggest that higher amounts of gypsum applied to lime-amended oxisols reduce the effectiveness of the main oxisol clay-sized mineral capable of adsorbing cations, i.e., kaolinite, to impair calcium leaching. The uptake data were complemented with some zeta-potential measurements, which supported the lack of substantial uptake of calcium even in the presence of sulfate. Some modeling calculations using the only available model covering sulfate and calcium on gibbsite have been done to rationalize the experimental data, but the model is only able to involve pure electrostatic attraction of calcium, which is not sufficient to produce substantial uptake. Finally, the aluminol basal plane that is present on both gibbsite and kaolinite has been additionally studied using second harmonic generation (SHG) down to 4°C, because the ion-pair formation decreases with decreasing temperature. The second harmonic results confirm the patterns observed in the electrokinetic measurements with kaolinite being quite comparable to the sapphire basal plane. Also and quite clearly, the presence of $CaSO_4$ solutions caused temperature dependence different from pure $CaCl_2$ and Na_2SO_4 solutions. The latter were essentially behaving like pure water. The difference between the calcium chloride and sulfate systems can be explained by sulfate interaction and might be linked to the temperature dependence of the formation of the $CaSO_4$ ion pair. The temperature dependency study could be an important starting point for looking at ice nucleation in the presence of the three different solutions and more strongly link aqueous chemistry to ice nucleation processes.

Keywords: acidic soils, double-layer screening, gypsum, leaching, oxisols, ice nucleation

1. Introduction

Gypsum (~95% m/m $CaSO_4.2H_2O$) can be used as soil amendment to lessen the phytotoxicity of available aluminum found in acidic subsoil depths that, although accessible to the plant roots, are not affected by surface liming [1]. Field and laboratory experiments have indicated that gypsum application on soil can increase the leaching of plant nutrients such as potassium, calcium, and magnesium [2–5]. Such an effect could be due to the formation of negative or neutral ion pairs (e.g., KSO_4^-, $CaSO_4^0$, and $MgSO_4^0$) in the soil solution as predicted by the hard and soft (Lewis) acid and base theory [6]. In contrast, experiments conducted with soil samples rich in gibbsite, hematite, goethite, or allophane have suggested that sulfate adsorption enhances calcium adsorption [7–13]. Fahrenhorst et al. [4] found that the gypsum potential of increasing the cation exchange capacity (CEC) of an oxisol from Amazon corresponded on a 1:1 basis with the sulfate adsorption by that soil. Okuma and Alves [14] observed, in the absence of phosphate, that the enhancing effect of sulfate on CEC increased with the soil contents of iron and aluminum oxides. Although Pearce and Sumner [11] have suggested that the equivalent retention of Ca^{2+} and SO_4^{2-} observed in kaolinitic subsoil could be due the combination of several mechanisms, including precipitation, specific adsorption, ionic strength-induced charging, and ion-pair adsorption, to date such phenomenon remains to be clarified.

Among the surface chemical equations used to model solution pH and ionic strength effects on ion complexation on a standard gibbsite surface ($\equiv AlOH^0$), those referring to sulfate complexation are given by [15]

$$\equiv AlOH^0 + SO_4^{2-} + H^+ \leftrightarrow \equiv AlSO_4^- + H_2O \quad (1)$$

$$\equiv AlOH^0 + SO_4^{2-} \leftrightarrow \equiv AlOHSO_4^{2-} \quad (2)$$

The above equations indicate that sulfate adsorption to gibbsite gives rise to negatively charged surface species, decreases the net positive surface charge, and increases the solution pH. Such features could favor sulfate-cation co-adsorption even at lower pH because sulfate adsorption on gibbsite increases as pH decreases [16]. Indeed, the acidic condition is required for the formation of ternary surface complexes of copper, lead, and cadmium with phosphate and ferrihydrite [17, 18]. Calcium complexation on gibbsite has in turn been modeled with the following equation [15]:

$$\equiv \text{AlOH}^0 + \text{Ca}^{2+} \leftrightarrow \equiv \text{AlOCa}^+ + \text{H}^+ \tag{3}$$

Although the surface chemical Eqs. (1)–(3) refer to adsorption sites located on the gibbsite edge faces, the basal planes of plate-shaped gibbsite and kaolinite as well seem to be operational in calcium adsorption as demonstrated by surface force measurements [19–21]. Considering that hydrolyzed species are preferably adsorbed to oxide surfaces [22], cations that hydrolyze at high pH values, such as calcium, would be expected not to be adsorbed on gibbsite and kaolinite edge sites at low pH. However, the above results obtained with atomic force microscopes indicate that cations presenting high hydrolysis constants can be retained on basal planes of gibbsite and kaolinite even under acidic condition.

Besides mineral surface properties, solution features must also be taken into account when evaluating ion adsorption. The formation of aqueous neutral ionic pairs can reduce the ion adsorption. The saturated solution of calcium sulfate (~15.8 mM), for instance, contains about 35% of total dissolved calcium and sulfate as the $CaSO_4^0$ ion-pair. Thus, both ions can be less adsorbed if such retention does not shift the equilibrium of $CaSO_4^0$ ion-pair formation toward the formation of free aqueous ions Ca^{2+} and SO_4^{2-}. Furthermore, even with limited aqueous ion-pair formation, an increase in the ionic strength can, in some cases, decrease ion adsorption on the solid phase [23, 24].

In this chapter, we studied the effects of the aqueous sulfate, salt concentration, and solution pH on the amounts of calcium adsorbed on kaolinite and gibbsite. These minerals were chosen because their joint contents in the oxisol clay fraction usually surpass 50% (m/m). Furthermore, while kaolinite is the most abundant cation-adsorbing mineral found in less weathered oxisols, gibbsite is the predominant clay-sized mineral in their deeply weathered counterparts [25]. The purpose of the present investigation was to evaluate the calcium adsorption behavior of kaolinite and gibbsite in oxisols amended with rich-sulfate sources such as gypsum.

2. Material and methods

2.1 Minerals

A kaolinite sample from Minas Gerais, Brazil, and a laboratory-prepared sample of gibbsite were used in the experiments. The gibbsite sample was prepared via titration of 500 mL of 1 M Al^{3+} (added as $AlCl_3.6H_2O$) with 315 mL of 4 M NaOH followed by precipitate dialysis for 1 month against daily refreshed deionized water kept at 50°C [26]. Both samples were washed with deionized water, oven-dried (75°C/48 h), ground in an agate mortar, and passed through a 0.3-mm sieve.

The mineralogical purity of the samples was checked through powder X-ray diffraction (XRD). The XRD patterns were measured using a Philips PW1830 diffractometer (PANalytical, Almelo, the Netherlands) in continuous scan mode (0.02°2θ s^{-1}) from 3 to 90°2θ CuKα. The particle shapes were examined via scanning electron microscopy (SEM) using a LEO 435 VP microscope (LEO Electron Microscopy Inc., Thornwood, NY). The pH at the isoelectric point (IEP) was determined by graphical interpolation from zeta-potential (ζ) measurements. The ζ values were measured in triplicate with a NanoBrook Omni analyzer (Brookhaven Instruments, Holtsville, NY) in mineral suspensions (1 g L^{-1}) prepared in 0.01 M KNO_3 solutions with different pH values, which were stabilized by additions of 0.01 M HCl or 0.01 M NaOH. The five-point N_2-BET-specific surface areas were measured in the samples with a Micromeritics ASAP 2010 surface area analyzer (Micromeritics, Norcross, GA).

2.2 Calcium adsorption experiments

Before the adsorption experiments, aqueous mineral suspensions (20 g L^{-1}) were titrated with 0.1 M HCl or 0.1 M NaOH until their pH values stabilized at 4 and 6 to mimic an unsuitable (pH 4) and a proper (pH 6) acidity condition for the most of crops. Then, the suspensions were centrifuged, and the minerals were oven-dried at 75°C for 48 h, ground in an agate mortar, and passed through a 0.3-mm sieve. Calcium solutions (0.5 and 10 mM; pH 4 and 6) were prepared from reagent-grade calcium sulfate and calcium nitrate. The above concentrations were chosen to allow for respective low and high ion pairing in the sulfate solutions. Aqueous calcium speciation (**Table 1**) was calculated using the formation constants available in the built-in database (NIST 46.7) of Visual MINTEQ [27].

Batch calcium adsorption experiments were performed without CO_2 exclusion as follows: 0.2 g of mineral and 20 mL of solution were placed into 50-mL polypropylene centrifuge tubes. The suspensions were shaken end over end at 30 rpm for 24 h at 20 ± 2°C. The pH was measured in the suspensions with a Thermo Orion 4-star pH meter (Thermo Orion Inc., Beverly, MA) after a two-point calibration with standard buffer solutions (pH 4.0 and 7.0). The tubes were centrifuged for 30 min at 10,000 rpm, and the supernatants were filtered through 0.22-μm cellulose membranes prior to calcium analysis in an atomic absorption spectrophotometer (Varian AA 240 FS, Agilent Technologies, Mulgrave, Australia). A cross-check of the initial calcium concentrations in the applied solutions was concurrent with the measurements performed in the remaining solutions. Both absolute and relative amounts of adsorbed calcium were calculated from the difference between the calcium initial and equilibrium aqueous concentrations.

Solution	pH	I^a	Ca^{2+}	$CaSO_4^0$	$CaNO_3^+$	$Ca^{2+\,c}$
		mM		%[b]		mM
0.5 mM $CaSO_4.2H_2O$	4.0	02	93.3	06.7	-	0.38
0.5 mM $CaSO_4.2H_2O$	6.0	02	93.2	06.2	-	0.38
0.5 mM $Ca(NO_3)_2.4H_2O$	4.0	02	99.7	-	0.3	0.42
0.5 mM $Ca(NO_3)_2.4H_2O$	6.0	02	99.7	-	0.3	0.42
10 mM $CaSO_4.2H_2O$	4.0	30	69.3	30.7	-	3.68
10 mM $CaSO_4.2H_2O$	6.0	30	69.2	30.8	-	3.68
10 mM $Ca(NO_3)_2.4H_2O$	4.0	30	96.8	-	3.2	5.07
10 mM $Ca(NO_3)_2.4H_2O$	6.0	30	96.8	-	3.2	5.07

[a]*$I = 1/2 \times \sum M_i z_i^2$, where I is the solution ionic strength calculated via Visual MINTEQ, mM; M_i and z_i are the respective concentration, mM, and charge of the dissolved specie i.*

[b]*Proportions of the main dissolved calcium species calculated via Visual MINTEQ in relation to the total calcium concentration.*

[c]*Free calcium activity in solution calculated via Visual MINTEQ from activity coefficients assessed with the Davies equation.*

Table 1.
Calcium solutions used in the sorption experiments.

2.3 Electrokinetic measurements

Electrokinetic experiments were carried out for kaolinite and gibbsite suspensions (1 g L^{-1}, pH 5) presenting increasing amounts of either calcium chloride or sodium sulfate. Additional measurements were carried out with calcium sulfate solutions including a saturated one (~15.8 mM) to maximize the formation of the aqueous $CaSO_4^0$ ion pair. The Brookhaven PALS apparatus was used. Suspensions were freshly prepared and at least 10 runs per sample were performed. These series clearly indicated the absence of time dependence, i.e., equilibration was typically on the order of 5 min.

2.4 Second harmonic generation measurements

Second harmonic generation (SHG) experiments were carried out with an alumina prism exposing the c-cut to the aqueous solution. This crystal plane is structurally equivalent to the gibbsite basal plane on both gibbsite and kaolinite. The setup has been described in detail elsewhere [28, 29]. The sapphire prism was obtained from Kyburz (Safnern, Switzerland) and cleaned based on procedures established earlier. The PTFE cell was initially cleaned as well with acetone and ethanol and copiously washed with MilliQ water. After each experimental series, the solution in the cell was replaced several times with MilliQ water until the signal obtained remained stable at the initial value of MilliQ water.

2.5 Data analysis

Uptake experiments were carried out in triplicate. The paired t-test was applied to evaluate if the mean difference between the amounts of calcium adsorbed to each mineral in the presence of nitrate and sulfate, respectively, differed from zero at $P = 0.05$. Calculations were performed using SAS software, version 9.3 [30].

The electrokinetic experiments involved at least 10 runs per sample. From this the averaged electrophoretic mobility was used, and the standard error was noted as specified by the software.

The second harmonic generation experiments allow clear observation of equilibrium states as well. As for the electrokinetic experiments, transient effects were absent when the reported data points were collected. The data were scaled to the signal in pure water, which was measured before and after a concentration series. The temperature dependence was studied with a setup that was previously used in ice-nucleation studies.

3. Results and discussion

3.1 Mineral characterization

The XRD patterns (**Figure 1**) show some accessory quartz with kaolinite and no other phase with gibbsite [31]. The plate-shaped morphologies of the mineral particles conform to the pseudohexagonal plates of kaolinite [32] and the plate-like lozenges or plate-like gibbsite hexagons [33] (**Figure 2**).

The specific surface area of the kaolinite sample (30.9 m^2 g^{-1}) is within the interval from 5 to 39 m^2 g^{-1} reported for this mineral [34]. The gibbsite surface area (26.6 m^2 g^{-1}) is close to that calculated by Rosenqvist and Casey [35] from the density and particle dimensions of a gibbsite sample prepared using the same

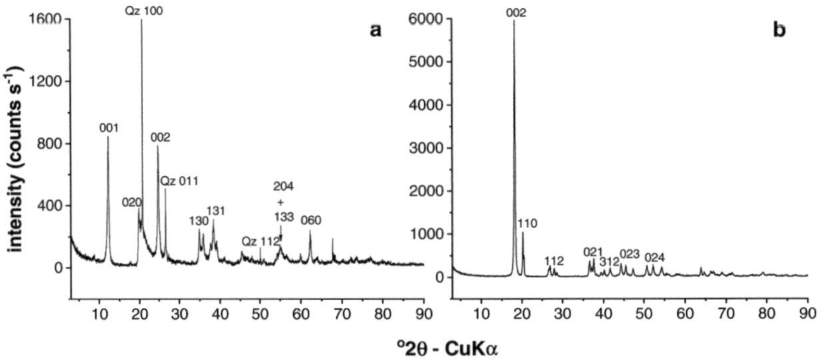

Figure 1.
X-ray diffraction patterns of (a) kaolinite and (b) gibbsite samples. Qz = quartz.

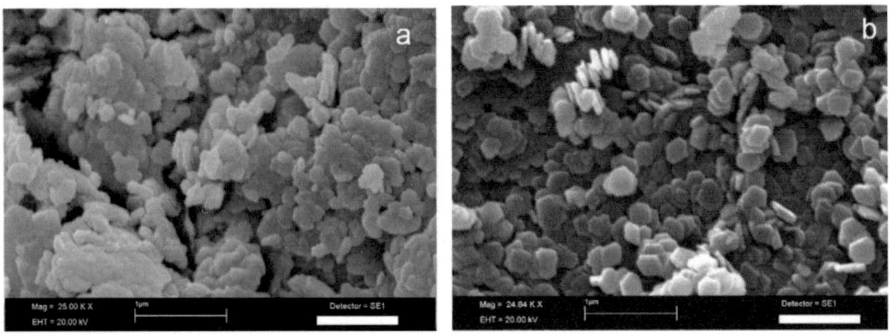

Figure 2.
Scanning electron micrographs of (a) kaolinite and (b) gibbsite samples.

procedure followed here (25 $m^2\ g^{-1}$). However, their reported N_2-BET surface area is somewhat lower (19.6 $m^2\ g^{-1}$).

The respective IEP values found for kaolinite and gibbsite are 4.9 and 10.6 (**Figure 3**). The kaolinite IEP is close to the value compiled by Stumm and Morgan [36] (4.6). The gibbsite one is in between the pristine point of zero charge at pH 10.0 found by Hiemstra et al. [26] and the IEP at pH 11.3 reported by Adekola et al. [37] for gibbsites with surface areas above 25 $m^2\ g^{-1}$.

3.2 Calcium adsorption on kaolinite

The batch experiments were carried out without attempting to exclude CO_2. In principle, calcite bulk precipitation could also decrease the equilibrium aqueous calcium concentrations. However, according to equilibrium calculations, the pH threshold value needed for calcite formation in the 10-mM calcium solutions in the absence of calcium adsorption is about 7.7. Considering that this pH value increases as the initial calcium concentration in solution decreases, calcite formation in our mineral suspensions can be safely excluded because the highest equilibrium pH was 6.9. Furthermore, we assumed that the aqueous bicarbonate (HCO_3^-) (~2.4 μM as calculated with Visual MINTEQ) did not affect the sulfate adsorption. Helyar et al. [38] have found that 0.58 mM HCO_3^- did not influence phosphate adsorption on a gibbsite sample.

The amounts of calcium adsorbed to kaolinite from the dilute and concentrated solutions are presented in **Table 2**. The results for the dilute solutions (0.5 mM) with initial pH 4 showed that higher sulfate adsorption on kaolinite at low pH [16] did not enhance calcium adsorption via potential co-adsorption. This finding

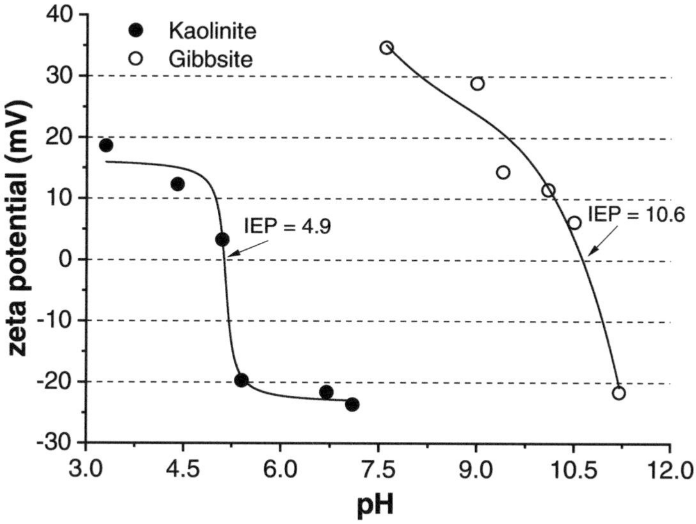

Figure 3.
Zeta potential values measured in kaolinite and gibbsite 0.01-M KNO_3 suspensions at different pH conditions. IEP = isoelectric point.

Solution	pH		Ca^{2+}	
	Initial	Equilibrium	$\mu mol\ m^{-2}$	(%)[a]
0.5 mM $Ca(NO_3)_2.4H_2O$	4.0	4.2	0.0a	00.0a
0.5 mM $CaSO_4.2H_2O$	4.0	4.4	0.0a	00.0a
0.5 mM $Ca(NO_3)_2.4H_2O$	6.0	5.1	0.4*b*	21.7*b*
0.5 mM $CaSO_4.2H_2O$	6.0	5.2	0.4*b*	22.7*b*
10 mM $Ca(NO_3)_2.4H_2O$	4.0	4.0	3.2A	09.4A
10 mM $CaSO_4.2H_2O$	4.0	4.1	1.6B	05.3B
10 mM $Ca(NO_3)_2.4H_2O$	6.0	4.7	1.6*A*	05.1*A*
10 mM $CaSO_4.2H_2O$	6.0	4.9	0.8*B*	02.6*B*

[a]*Calculated in relation to the initial calcium concentration.*

Means followed by the same letter in the same column do not differ at P = 0.05 according to the paired t-test. Lowercase letters refer to comparisons of dilute solutions (0.5 mM); uppercase letters refer to comparisons of concentrated solutions (10 mM).
Normal letters refer to comparisons between solutions at initial pH 4; italicized letters refer to comparisons between solutions at initial pH 6.

Table 2.
Initial and equilibrium pH values and amounts of calcium adsorbed on kaolinite in the presence of nitrate and sulfate.

suggests that a possible increase in negative surface charges following from sulfate adsorption was insufficient to overcome the electrostatic repulsion to calcium ions exerted by positive surface charges. In this case, positive charges superseded negative ones because the equilibrium pH values remained below the kaolinite IEP. These results also indicate that even starting to deprotonate at pH 3 [39], the silanol surface groups of kaolinite did not adsorb calcium. The local electrostatic repulsion from the positively charged aluminol surface groups of kaolinite, which deprotonate above pH 8.7 [39], may have outweighed the attraction exerted by the negatively charged silanol groups on calcium at low pH and low ionic strength. Calcium adsorption to kaolinite from the 0.5-mM solutions with initial pH 6 was the same in the presence of either nitrate or sulfate. Therefore, no sulfate-calcium co-adsorption occurred under the net negative surface charge condition provided by equilibrium pH values (5.1 and 5.2) higher than the kaolinite IEP (4.9).

The increase in the initial concentration of calcium sulfate (10 mM) reduced calcium adsorption on kaolinite by 50% relative to those measured in the presence of 20 $mmol_c\ L^{-1}$ of nitrate for the two initial pH conditions. Equilibrium calculations (**Table 1**) indicate that the neutral pair $CaSO_4^0$ comprises 31% of the total calcium dissolved in the 10 mM calcium sulfate solution for both pH values, which decreases free calcium ion activity in the sulfate solutions to about 63% of those calculated for the nitrate solutions. The enhanced formation of neutral ion pairs containing an adsorptive and/or lowering the activity of its ionic free form in solution may decrease its adsorption [23, 40] and outweigh potential co-adsorption. Furthermore, the reduction of Ca^{2+} and SO_4^{2-} activities in solution due to adsorption did not seem to promote an appreciable shift in the equilibrium of $CaSO_4^0$ formation toward free aqueous Ca^{2+} and SO_4^{2-}, which could lead to comparable conditions of free ions in the nitrate solutions. Finally, unlike the observed for the dilute solutions presenting initial pH 4, the enhancement of calcium loading to 10 mM resulted in calcium adsorption on kaolinite overcoming the electrostatic repulsion from the net positive surface charge of that mineral at equilibrium.

3.3 Calcium adsorption on gibbsite

The amounts of calcium adsorbed on gibbsite from all solutions are presented in **Table 3**. Although according to Eqs. (1) and (2) sulfate adsorption to gibbsite could favor sulfate-calcium coadsorption, such a trend was not observed for the dilute and acidic conditions (0.5 mM; pH 4). At equilibrium pH ~ 7, calcium adsorption on gibbsite from the dilute solutions only occurred in the presence of sulfate. Because of the high IEP of the studied gibbsite, all adsorption experiments corresponded to a net positive surface charge. Sulfate adsorption and pH enhancement reduce the positive charges and concomitant electrostatic repulsion of calcium. Therefore,

Solution	pH		Ca^{2+}	
	Initial	Equilibrium	$\mu mol\ m^{-2}$	(%)[a]
0.5 mM $Ca(NO_3)_2.4H_2O$	4.0	4.3	0.0a	0.0a
0.5 mM $CaSO_4.2H_2O$	4.0	4.5	0.0a	0.0a
0.5 mM $Ca(NO_3)_2.4H_2O$	6.0	6.5	0.0*a*	0.0*a*
0.5 mM $CaSO_4.2H_2O$	6.0	6.9	0.2*b*	8.2*b*
10 mM $Ca(NO_3)_2.4H_2O$	4.0	4.4	1.9A	4.3A
10 mM $CaSO_4.2H_2O$	4.0	4.6	0.9B	2.6B
10 mM $Ca(NO_3)_2.4H_2O$	6.0	6.2	0.0*A*	0.0*A*
10 mM $CaSO_4.2H_2O$	6.0	6.8	0.0*B*	0.0*B*

[a]*Calculated in relation to the initial calcium concentration.*

Means followed by the same letter in the same column do not differ at P = 0.05 according to the paired t-test. Lowercase letters refer to comparisons of dilute solutions (0.5 mM); uppercase letters refer to comparisons of concentrated solutions (10 mM).
Normal letters refer to comparisons between solutions at initial pH 4; italicized letters refer to comparisons between solutions at initial pH 6.

Table 3.
Initial and equilibrium pH values and amounts of calcium adsorbed on gibbsite in the presence of nitrate and sulfate.

our results suggest that besides sulfate adsorption, the net positive surface charge, which also depends on solution pH and ionic strength, affected calcium sulfate co-adsorption on gibbsite. Sulfate also decreased calcium adsorption on gibbsite from the more concentrated acidic solutions (10 mM; pH = 4), presumably due to the same mechanisms as those proposed for kaolinite.

4. pH and salt concentration effects on calcium adsorption on kaolinite and gibbsite

Unlike the enhancing pH effect on calcium adsorption on kaolinite observed for the 0.5-mM solutions containing either nitrate or sulfate, an opposite effect was found for the two more concentrated solutions (10 mM) containing those anions. In these cases, an initial pH increase from 4 to 6 decreased absolute calcium adsorption by 50% on kaolinite and by 100% on gibbsite.

This negative pH effect on calcium adsorption from the 10-mM solutions can be considered analogous to the positive pH effect on orthophosphate adsorption on kaolinite observed in previous papers for acidic condition (pH < 7) [41–43]. Such an effect concurs with negative charge enhancement on the kaolinite surface, which disfavors orthophosphate adsorption. On the other hand, the orthophosphate adsorption to kaolinite from dilute solutions decreases as solution pH increases, which resembles the P adsorption behavior of iron oxides [44]. He et al. [45] suggested that the aqueous P speciation may be related to the positive pH effect on orthophosphate adsorption to kaolinite. However, such an association does not seem plausible: according to equilibrium calculations based on NIST 46.7 stability constant database, the aqueous orthophosphate species ($H_2PO_4^-$) prevails from pH ranging from 3 to 7. Likewise, for pH ranging from 4 to 7, calcium speciation in the studied 10-mM solutions does not differ from results in **Table 1**. Therefore, higher salt concentration should be associated with low calcium adsorption at higher pH values. This hypothesis will be discussed below.

While the basal planes of kaolinite have been considered to hold permanent negative charges [46], recent surface force measurements carried out with atomic force microscopes have shown that the surface charges of the silica basal plane of kaolinite and the alumina basal plane of kaolinite and gibbsite react to changes in solution pH and salt concentration as the variable charges found on edge faces of those minerals [19–21]. Furthermore, the solution effects on basal silica surface charges differ from those observed for alumina faces [19, 21], and even for a given basal plane, the magnitudes of the surface charges can change when the dissolved cation constitutes the sole difference among the solutions in contact with the mineral [20].

Such complex, anisotropic charge behavior may be relevant to our findings given the apparent unexpected negative pH effect on calcium adsorption. Weak calcium hydrolysis suggests that uptake on oxide-like surface groups, such as those on kaolinite and gibbsite edges, only occurs at relatively high pH [15, 47]. Therefore, the basal planes of kaolinite and gibbsite may be the main adsorption sites of that cation at pH < 7 [19, 21]. Although Siretanu et al. [20] observed calcium adsorption from $CaCl_2$ solutions at pH 6 for the alumina face of nanosized gibbsite, their surface force measurements indicated that calcium adsorption from solutions with increasing concentrations of that cation initially increased followed by a decrease in the extracted charge densities with a maximum between 5 and 10 mM $CaCl_2$. The authors argued that the concurrent increasing co-adsorption of chloride ions could explain the decrease in surface force. Unfortunately, the amounts of adsorbed calcium cannot be assessed in such experiments as from batch adsorption studies. On the other hand, batch adsorption results cannot be associated with a given surface plane. The maximum diffuse layer potential from force measurements can likewise be explicated by constant or increasing calcium adsorption overcompensated by screening due to chloride adsorption in the Stern layer or simple double-layer screening.

In our experiments with the 10-mM solutions, the mean potential differences across the electrical double layer of gibbsite at equilibrium, as estimated by the Nernst Equation [48], were + 360 and + 242 mV for the more and less acidic solutions, respectively. These values, which are upper limit estimates, indicate that anion adsorption should be disfavored in the less acidic suspension. In the kaolinite scenarios with the 10-mM solutions, the mean Nernst surface potentials were also positive at equilibrium, i.e., +50 and + 5.6 mV for more and less acidic solutions, respectively. The lower values compared to those calculated for gibbsite, and a possible difference between the screening effects on silica and alumina basal planes, could cause a moderate reduction in calcium adsorption on kaolinite from the less acidic nitrate and sulfate solutions. For the experiments with the dilute solutions (0.5 mM), the low salt concentration was insufficient for promoting appreciable screening effects.

4.1 Electrokinetic measurements

Figure 4 displays the results of electrokinetic measurements carried out for kaolinite and gibbsite suspensions at pH 5. In general, the electrokinetic mobility (EM) values measured against increasing salt concentrations showed the same trends for both studied minerals. The increasing sulfate addition as Na_2SO_4 enhanced the negative surface charge of both kaolinite and gibbsite (**Figure 4(a)**). However, in the gibbsite scenario, sulfate caused charge reversal at the concentration of about 0.3 mM Na_2SO_4. Data obtained via the increasing addition of $CaCl_2$ to kaolinite and gibbsite in the presence of 3 mM Na_2SO_4 demonstrated rather a shielding effect, mainly for the gibbsite scenario where no clear charge reversal was observed at 3 mM Ca^{2+} (**Figure 4(b)**). In a mixed $CaSO_4$ system up to the solubility limit, we retrieved the sulfate effect of increasing the surface negative charge (**Figure 4(c)**). Electrophoretic mobility measurements are usually unstable when performed at pH value close to the mineral IEP; such an instability probably resulted in EM values somewhat different for kaolinite in the absence of Na_2SO_4 (**Figure 4(a)**) and $CaSO_4$ (**Figure 4(c)**). Despite this, the obtained results corroborate the idea that co-adsorption of calcium due to sulfate adsorption is a minor feature.

4.2 Surface charge modeling applied to gibbsite

Some of our assumptions also agree with gibbsite surface charge trends simulated with the diffuse layer model (DLM) [49] using the calcium/gibbsite and sulfate/gibbsite surface complexation constants available in [15] as well as Eqs. (1)–(3). The simulations were performed for standard gibbsite suspensions (10 g L^{-1}) in solutions containing dual cation-anion combinations with sodium, calcium, nitrate, and sulfate, where nitrate and sodium ions do not interact with the surfaces via chemical reactions. Binary systems were chosen as references of both inert ions (sodium and nitrate) and their mixtures with sulfate and calcium. Although the standard gibbsite as treated by Karamalidis and Dzombak [15] has an IEP of 9 (i.e., lower than in the present case), the calculation results illuminate calcium adsorption in the mixed system. Unfortunately, self-consistent data for a more complex surface complexation model are not available. The authors are well

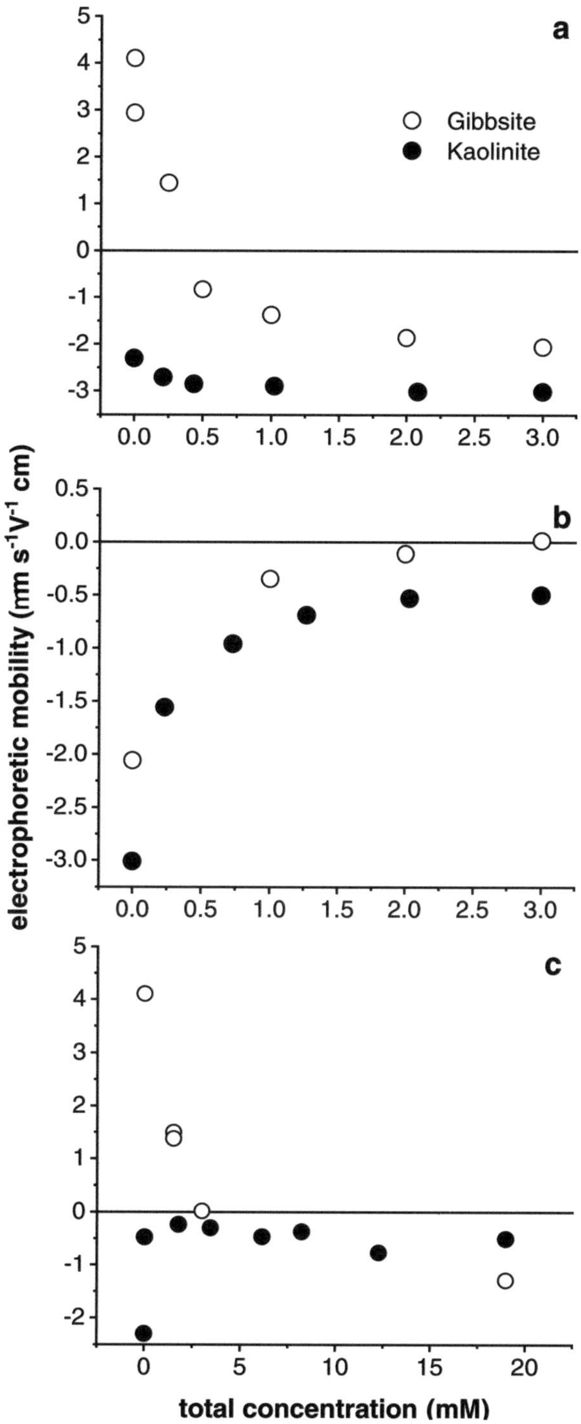

Figure 4.
Electrophoretic mobility values of kaolinite and gibbsite suspensions measured under increasing concentrations of (a) Na_2SO_4, (b) $CaCl_2$ + 3 mM Na_2SO_4, and (c) $CaSO_4$.

aware about the drawback of using a simple DLM. With the available data, it was not possible to design a separate model. The present modeling, for example, disregards the available information about the nature of the surface complexes (inner sphere vs. outer sphere) or the calcium adsorption on the basal planes, because the

DLM only includes inner-sphere surface complexes and the model does not distinguish crystal planes. Also when the data base was generated, the information about the Ca adsorption at low pH had not been available. Furthermore, in the following the comparisons between solutions containing 1:1, 2:1 and 2:2 electrolyte solutions are difficult to directly compare, since at the same time the ionic strength and the cation concentrations might be changing for a given concentration of the electrolyte (this is also true for the subsequent sections).

The enhancement of an inert electrolyte concentration increases the gibbsite charge at a fixed pH with no shift in the IEP (**Figure 5(a)**). The diffuse layer model in the database used for the calculations causes a steep charge increase due to the absence of a Stern layer. This behavior is hidden when plotting titration raw data (i.e., total acid and base added versus pH). As depicted in **Figure 5(b)**, calcium does not significantly affect the curves relative to **Figure 5(a)**. The charge for a given cation concentration and pH increases because the negative counterion concentration is higher in the calcium presence, allowing for better shielding and increased charge. The enhanced charge is not due to calcium adsorption, which is insignificant.

Figure 5(c) shows two effects relative to the gibbsite surface charge as a function of sodium sulfate concentration. One is sulfate uptake, which greatly decreases surface charge density. For the two lowest sulfate concentrations, all sulfate is adsorbed with nearly no negatively charged counterion to shield positive charges, thereby limiting the number of positively charged surface groups. The situation changes for the high-sulfate concentration: first, the IEP is decreased due to adsorbed sulfate. Less than 15% sulfate is adsorbed at the lowest pH, leaving a substantial negative charge to allow the increase in protonated surface groups as pH decreases. Calcium sulfate (**Figure 5(d)**) behaves similarly to sodium sulfate; according to the model, calcium adsorption does not occur at low pH. The formation of the $CaSO_4^0$ ion pair in solution affects ionic strength and free sulfate activity; hence the charge is higher in the sodium sulfate system. The model does not predict enhanced calcium uptake and indicates that no system charge reversal occurred at the pH values investigated experimentally. Even so, a possible relative increase in neutral aluminol groups ($\equiv AlOH^0$) at pH ~ 7 occurred in our experiments to favor the sulfate-calcium co-adsorption on gibbsite from the 0.5 mM $CaSO_4$ solution.

4.3 Considerations from kaolinite titration data

Riese [50] titrated kaolinite in $CaCl_2$, $NaNO_3$, and Na_2SO_4 solutions, and we show the resulting curves in **Figure 6**. The results for $NaNO_3$ (**Figure 6(a)**) showed a typical trend with increasing salt concentration, i.e., a higher titrable charge in a higher electrolyte solution. The absence of spread as a function of salt level observed at low Na_2SO_4 concentrations (**Figure 6(b)**) could indicate sulfate adsorption when related to the calculations for gibbsite (**Figure 5(c)**). However, for the previous gibbsite charge simulations, the total charge was assessed, whereas for kaolinite the titrable one was determined. The titration data in $CaCl_2$ showed a very small spread and surprisingly low values for the titrable charge as compared to $NaNO_3$ (**Figure 6(c)**) perhaps because the interaction of calcium with the kaolinite surface hindered proton release. Such an interaction would not apply to the kaolinite edge; also, oxidic surface functional groups on edge faces are not expected to bind to calcium at low pH values. Thus, the interaction is more likely with basal planes. Kumar et al. [21] found similar features on the kaolinite gibbsite face and reported visible co-adsorption of chloride at 100 mM $CaCl_2$ concentration. The influence of calcium adsorption on basal surfaces on proton release must be studied to understand the systems. Clearly, the available data suggest that both sulfate and

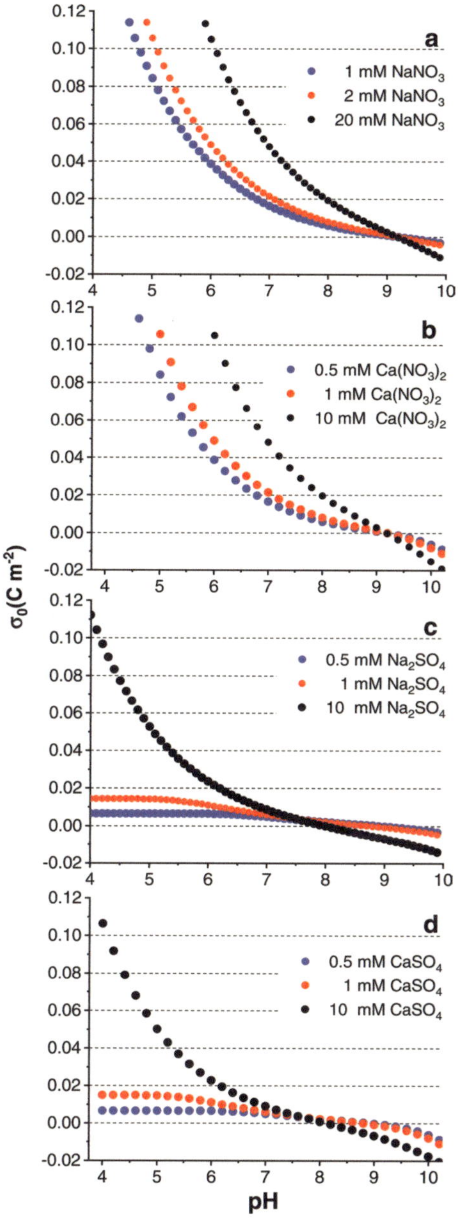

Figure 5.
Net surface charge (σ_o) values calculated using the double layer model for standard gibbsite in 10-g L^{-1} aqueous suspensions as a function of pH, salt concentration, and composition: (a) $NaNO_3$, (b) $Ca(NO_3)_2$, (c) Na_2SO_4, and (d) $CaSO_4$.

calcium each interact with kaolinite at pH 4 and 6, rendering cooperative interactions possible; this trend supports experimental data from our $CaSO_4$ systems.

4.4 Second harmonic generation experiments at room temperature

Figure 7 shows results of second harmonic generation (SHG) experiments at pH 6 with the c-cut of sapphire. The setup has been described in much detail before [28, 29]. As pointed out before, this crystal plane is more or less structurally equivalent to the gibbsite basal plane, which is present on both gibbsite and kaolinite. Numerous investigations have shown that this plane has a low isoelectric point.

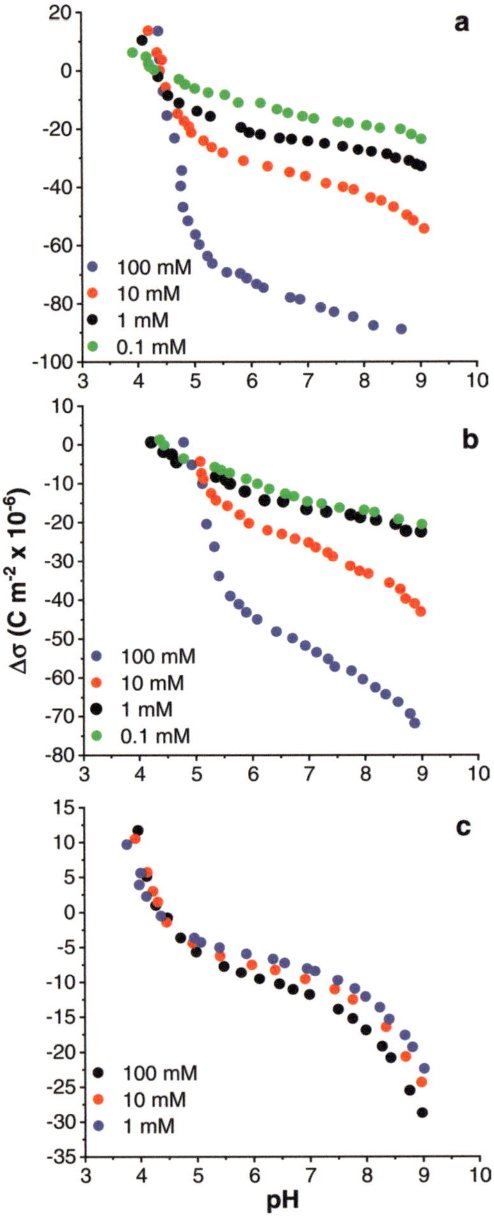

Figure 6.
Titrable surface charge (Δσ) values obtained by Riese [50] for kaolinite KGa-1 suspended in solutions of (a) $NaNO_3$, (b) Na_2SO_4, and (c) $CaCl_2$.

So it is probable that at pH 6, the surface is negatively charged. **Figure 7(a)** shows the results for Na_2SO_4 addition that correspond to those shown in **Figure 4(a)**. The experiment is started from a water solution to which Na_2SO_4 is added, and the signal is referenced to the signal in water (this is also the case for all the other data in **Figure 7**). The increase in the SHG signal indicates less positive or more negative charge. Assuming that the sapphire surface is negative at this pH value, it would mean more negative charge and correspond to data on kaolinite rather than those on gibbsite. Also the change in signal is quite small, which also is the case for kaolinite in **Figure 4(a)**. The unexpected adsorption of anions on the negatively charged basal plane of sapphire has also been reported by others. Starting from a solution containing about 1.5 mM sulfate, the addition of $CaCl_2$ results in a much

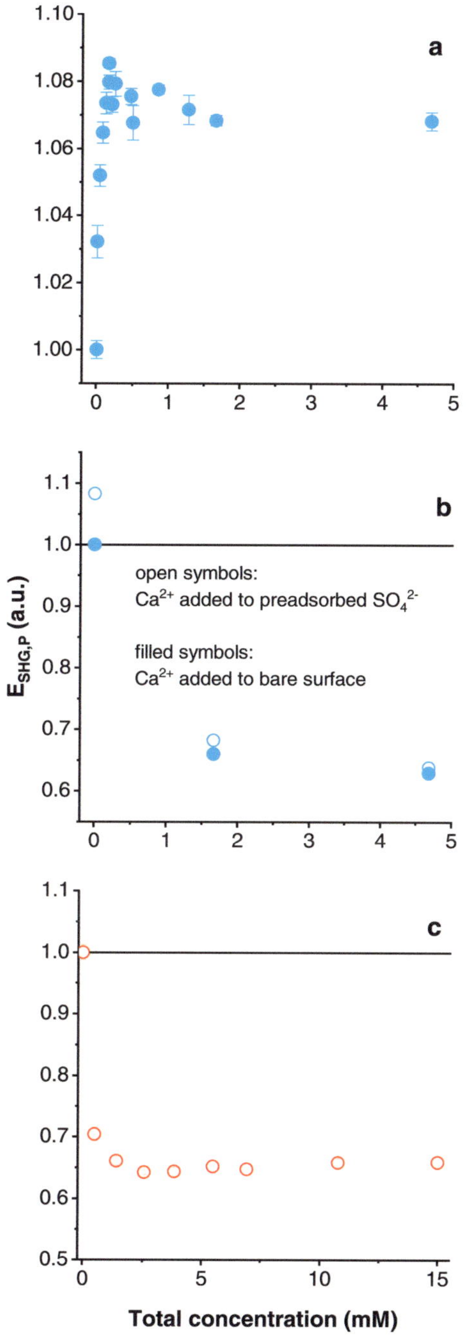

Figure 7.
Second harmonic generation electric field of sapphire basal planes measured under increasing concentrations of (a) Na_2SO_4, (b) $CaCl_2$ + 1.5 mM Na_2SO_4, and (c) $CaSO_4$.

stronger change in the signal (**Figure 7(b)**). This is the same as what was observed in **Figure 4(b)** for both gibbsite and kaolinite. The SHG results on the basal plane clearly mimic the electrokinetic behavior. Furthermore, addition of $CaCl_2$ to the bare surface shows the same behavior irrespective of the absence and presence of sulfate as can be seen in **Figure 7(b)**. This would suggest that the divalent cation is strongly shielding the negative charge, whereas it is not clear whether charge reversal occurs. The SHG data cannot indicate charge reversal. The electrokinetic data for gibbsite might suggest charge reversal at higher Ca concentrations, which in turn would agree with the observation of Siretanu et al. [20] who directly observed the specific interaction of Ca with gibbsite basal planes. As for **Figure 7(a)**, the results for the basal plane from **Figure 7(b)** concur with the data for the particles from **Figure 4(b)**, which could suggest the relevance of the basal planes in the reactions of the particles. Finally in **Figure 7(c)**, $CaSO_4$ solutions have been added to the sapphire basal plane to mimic the results from **Figure 4(c)**. Similar to the discussion with respect to **Figures 4(a)** and **7(a)**, it appears that the SHG is more comparable to the kaolinite system, since the decrease of the signal is rather showing either an increase in positive charge or a decrease in negative charge. The latter happens in **Figure 4(a)** for kaolinite, but none of the options would be applicable for gibbsite. Also the strong change at the lower concentration and the quick plateauing occurs for the kaolinite (**Figure 4(c)**) and for the sapphire basal plane (**Figure 7(c)**). While this cannot exclude a role played by sulfate that is very apparent for gibbsite (**Figure 4(c)**), the predominant role of calcium is obvious. So the role of the calcium sulfate ion pair remains elusive also based on these results, while the small but discernible increase in the signal in the $CaSO_4$ system indicates sulfate interaction at concentrations above 5 mM (**Figure 7(c)**).

To further investigate the role of these ion pairs, temperature-dependent experiments were carried out. The results will be presented and discussed in the final section.

4.5 Second harmonic generation experiments at variable temperature

Preliminary speciation calculations using various data bases were performed to gain knowledge about the extent of ion-pair formation when the temperature is changed. Calculations were done for 15 mM solutions of divalent ions, starting from 20°C. **Figure 8(a)** shows that with decreasing temperature the extent of ion-pair formation decreases (for $CaCl_2$) or remains approximately constant (for Na_2SO_4). For both systems the dominant ion pairs make up less than 10%. This is different for the $CaSO_4$ system, where (as could be expected) a decrease in the extent of ion-pair formation occurs with decreasing temperature. Clearly also the ion pair makes up more than 30% as discussed in previous sections. The concomitant solutions were put into contact with sapphire basal planes and studied as a function of temperature by second harmonic generation. The setup used in this case has been extensively described [28, 29]. **Figure 8(b)** shows results of second harmonic generation (SHG) experiments, starting with 15 mM solutions of the divalent ions at pH 6 and room temperature with decreasing temperature. In water the SHG signal is changing little. This is similar for the sodium sulfate system, which is also similar to water. In agreement with the results shown in **Figure 7(a)**, where only small changes were observed in that system. Actually data not shown for about 15 mM sulfate showed on average 1.06 times the water signal. With the average water signal at 0.535 at 20°C, we obtain a value of 0.567 in the presence of 15 mM sulfate, which is within the range of signals collected in the temperature dependence curve. The black curve also suggests a slight decrease of the signal for the sulfate signal compared to water at 4°C, but even here the scatter remains too large to draw ultimate conclusions. With decreasing temperature the sulfate adsorption should increase, but with respect to **Figure 7**, it was concluded that Ca was the more important ion in this system. The more interesting systems in this temperature range in the SHG experiments with the c-cut of sapphire are clearly also the Ca systems. The quite strong decrease in the signal for both Ca systems is observed. The stronger extent in $CaCl_2$ solutions at 15 mM is also observed (giving 0.6 of the water signal, data now shown). Most interestingly, a trend with temperature occurs in the chloride system

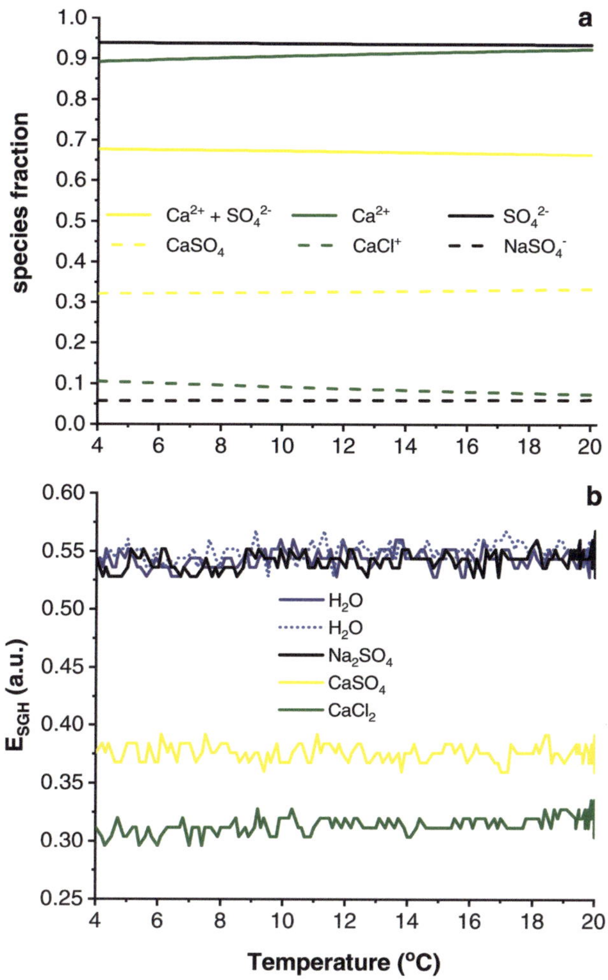

Figure 8.
Comparison of Na_2SO_4 (black) $CaCl_2$ (green) $CaSO_4$ (yellow) solutions (15 mM of divalent ions at 20°C) between 4 and 20°C with respect to speciation (a) and second harmonic signal (b). (b) also includes two runs with water only, one before and one after the experiments with the salt solutions.

to lower signal with lower temperature, while the $CaSO_4$ system remains constant. If a decrease in the signal is associated with stronger positive charge, it would suggest that there is a role of sulfate in the $CaSO_4$ system, which may be triggered by the decreased ion-pair formation, making more sulfate available with decreasing temperature. Note that overall much more free Ca is available in the $CaCl_2$ system, even if it is slightly decreasing with decreasing temperature. So decreasing temperature would favor at the same time enhanced sulfate interaction (anions adsorb more strongly on oxides with decreasing temperature) and make more free ions available, which would also favor uptake. Based on the results, the effects of temperature on sulfate would then outcompete those on calcium, but the role of the ion-pair formation could become visible. More experiments with different concentrations would be helpful in gaining more insight. Additional experiments with more soluble $MgSO_4$ are also planned. The formation of the ion pairs can also be an issue in freezing processes on surfaces in the presence of ions like Ca and sulfate, which can be expected in aerosols due to their presence in sea water. This could be an interesting link between the aqueous chemistry and the atmospheric chemistry which has not been explored to the extent possible.

5. Conclusions

The findings of this chapter, which are associated to acidity conditions ranging from an unsuitable to a more favorable one for cropping, indicate that sulfate adsorption has no or little positive effect on calcium adsorption on kaolinite and gibbsite. On the other hand, as both calcium and sulfate concentrations in the solution phase are increased toward the precipitation threshold of calcium sulfate, calcium adsorption on those minerals is inhibited. Further, for a higher initial calcium loading (10 mM), the equilibrium pH enhance from 4.0 to 6.9 decreases calcium adsorption to kaolinite and gibbsite. Besides indicating low calcium adsorption on gibbsite, the experimental data suggest that the application of higher amounts of sulfate-concentrated products such as gypsum on lime-amended oxisols tends to reduce the effectiveness of kaolinite, the main clay-sized mineral responsible for the cation exchange capacity of those soils, to impair calcium leaching. The results also suggest that hematite and goethite are the main oxisol clay-sized minerals responsible for the calcium sulfate co-adsorption reported to oxisol samples. This hypothesis should be evaluated in future research. From the data presented in this chapter, it seems that sulfate at the pH conditions considered dominates the gibbsite behavior, while on kaolinite the calcium is the more relevant ion. The basal plane of sapphire as studied with second harmonic generation yields results that are in close agreement with those of the kaolinite particles, which suggests that the kaolinite behavior might be explained by its gibbsite plane. The role of the ion pair in solution and of sulfate even on negatively charge surfaces has been discussed, and the relevance of studying these features also with respect to atmospheric sciences (freezing of water and ice nucleation on particle surface) was addressed. In particular the link between aqueous chemistry and ice nucleation should be used to understand processes the atmosphere.

Author details

Ahmed Abdelmonem[1*], Yujun Wang[2], Johannes Lützenkirchen[3] and Marcelo Eduardo Alves[4]

1 Institute of Meteorology and Climate Research, Karlsruhe Institute of Technology, Karlsruhe, Germany

2 Key Laboratory of Soil Environment and Pollution Remediation, Institute of Soil Science, Chinese Academy of Sciences, Nanjing, PR China

3 Institute for Nuclear Waste Disposal, Karlsruhe Institute of Technology, Karlsruhe, Germany

4 Departamento de Ciências Exatas, Escola Superior de Agricultura 'Luiz de Queiroz' – ESALQ/USP, Piracicaba, SP, Brasil

*Address all correspondence to: ahmed.abdelmonem@kit.edu

References

[1] Van Raij B. Gesso na Agricultura. Campinas: Instituto Agronômico; 2008. p. 233

[2] Pavan MA, Bingham FT, Pratt PF. Redistribution of exchangeable calcium, magnesium, and aluminum following lime or gypsum applications to a Brazilian oxisol. Soil Science Society of America Journal. 1984;**48**:33-38. DOI: 10.2136/sssaj1984.03615995004800010006x

[3] Alva AK, Gascho GJ. Differential leaching of cations and sulfate in gypsum amended soils. Communications in Soil Science and Plant Analysis. 1991;**22**:1195-1206. DOI: 10.1080/00103629109368484

[4] Fahrenhorst C, Botschek J, Skowronek A, Ferraz J. Application of gypsum and lime to increase cation adsorption of a Geric Ferralsol in the Brazilian Amazon region. Journal of Plant Nutrition and Soil Science. 1999;**162**:41-47. DOI: 10.1002/(SICI)1522-2624(199901)162:1<41::AID-JPLN41>3.0.CO;2-6

[5] Ernani PR, Miquelluti DJ, Fontoura SMV, Kaminski J, Almeida JA. Downward movement of soil cations in highly weathered soils caused by addition of gypsum. Communications in Soil Science and Plant Analysis. 2006;**37**:571-586. DOI: 10.1080/00103620500449443

[6] Pearson RG. Hard and soft acids and bases. Journal of the American Chemical Society. 1963;**85**:3533-3539. DOI: 10.1021/ja00905a001

[7] Bolan NS, Syers JK, Sumner ME. Calcium-induced sulfate adsorption by soils. Soil Science Society of America Journal. 1993;**57**:691-696. DOI: 10.2136/sssaj1993.03615995005700030011x

[8] Marcano-Martinez E, Mcbride MB. Calcium and sulfate retention by 2 Oxisols of the Brazilian cerrado. Soil Science Society of America Journal. 1989;**53**:63-69. DOI: 10.2136/sssaj1989.03615995005300010012x

[9] Alva AK, Sumner ME, Miller WP. Reactions of gypsum or phosphogypsum in highly weathered acid subsoils. Soil Science Society of America Journal. 1990;**54**:993-998. DOI: 10.2136/sssaj1990.03615995005400040010x

[10] Davis JG, Burgoa B. Interactive mechanisms of anion adsorption with calcium leaching and exchange. Soil Science. 1995;**160**:256-264. DOI: 10.1097/00010694-199510000-00004

[11] Pearce RC, Sumner ME. Apparent salt sorption reactions in an unfertilized acid subsoil. Soil Science Society of America Journal. 1997;**61**:765-772. DOI: 10.2136/sssaj1997.03615995006100030009x

[12] Cichota R, Vogeler I, Bolan NS, Clothier BE. Cation influence on sulfate leaching in allophanic soils. Australian Journal of Soil Research. 2007;**45**:49-54. DOI: 10.1071/SR06070

[13] Cichota R, Vogeler I, Bolan NS, Clothier BE. Simultaneous adsorption of calcium and sulfate and its effect on their movement. Soil Science Society of America Journal. 2007;**71**:703-710. DOI: 10.2136/sssaj2006.0206

[14] Okuma DM, Alves ME. Anion and mineralogical effects on K^+, Ca^{2+}, and Mg^{2+} leaching in oxisols. Soil Science. 2011;**176**:115-123. DOI: 10.1097/SS.0b013e31820efe4c

[15] Karamalidis AK, Dzombak DA. Surface Complexation Modeling: Gibbsite. Hoboken: John Wiley & Sons Inc.; 2010. p. 294

[16] Essington ME, Anderson RM. Competitive adsorption of 2-ketogluconate and inorganic ligands onto gibbsite and kaolinite. Soil Science Society of America Journal. 2008;**72**:595-604. DOI: 10.2136/sssaj2007.0190

[17] Tiberg C, Sjostedt C, Persson I, Gustafsson JP. Phosphate effects on copper(II) and lead(II) sorption to ferrihydrite. Geochimica et Cosmochimica Acta. 2013;**120**:140-157. DOI: 10.1016/j.gca.2013.06.012

[18] Tiberg C, Gustafsson JP. Phosphate effects on cadmium(II) sorption to ferrihydrite. Journal of Colloid and Interface Science. 2016;**471**:103-111. DOI: 10.1016/j.jcis.2016.03.016

[19] Liu J, Sandaklie-Nikolova L, Wang XM, Miller JD. Surface force measurements at kaolinite edge surfaces using atomic force microscopy. Journal of Colloid and Interface Science. 2014;**420**:35-40. DOI: 10.1016/j.jcis.2013.12.053

[20] Siretanu I, Ebeling D, Andersson MP, Stipp SLS, Philipse A, Stuart MC, et al. Direct observation of ionic structure at solid-liquid interfaces: A deep look into the Stern layer. Scientific Reports. 2014;**4**:4956. DOI: 10.1038/Srep04956

[21] Kumar N, Andersson MP, Van den Ende D, Mugele F, Siretanu I. Probing the surface charge on the basal planes of kaolinite particles with high-resolution atomic force microscopy. Langmuir. 2017;**33**:14226-14237. DOI: 10.1021/acs.langmuir.7b03153

[22] Stumm W, Hohl H, Dalang F. Interaction of metal-ions with hydrous oxide surfaces. Croatica Chemica Acta. 1976;**48**:491-504

[23] Mattigod SV, Gibali AS, Page AL. Effect of ionic-strength and ion-pair formation on the adsorption of nickel by kaolinite. Clays and Clay Minerals. 1979;**27**:411-416. DOI: 10.1346/Ccmn.1979.0270603

[24] Hayes KF, Leckie JO. Modeling ionic-strength effects on cation adsorption at hydrous oxide-solution interfaces. Journal of Colloid and Interface Science. 1987;**115**:564-572. DOI: 10.1016/0021-9797(87)90078-6

[25] Alves ME, Omotoso O. Improving Rietveld-based clay mineralogic quantification of oxisols using Siroquant. Soil Science Society of America Journal. 2009;**73**:2191-2197. DOI: 10.2136/sssaj2008.0365

[26] Hiemstra T, Yong H, Van Riemsdijk WH. Interfacial charging phenomena of aluminum (hydr)oxides. Langmuir. 1999;**15**:5942-5955. DOI: 10.1021/la981301d

[27] Gustafsson JP. Visual MINTEQ. Version 3.1. Stockholm: KTH Royal Institute of Technology; 2018. Available at: http://vminteq.lwr.kth.se

[28] Abdelmonem A, Lützenkirchen J, Leisner T. Probing ice-nucleation processes on the molecular level using second harmonic generation spectroscopy. Atmospheric Measurement Techniques. 2015;**8**:3519-3526. DOI: 10.5194/amt-8-3519-2015

[29] Lützenkirchen J, Abdelmonem A, Weerasooriya R, Heberling F, Metz V, Marsac R. Adsorption of dissolved aluminum on sapphire-c and kaolinite: Implications for points of zero charge of clay minerals. Geochemical Transactions. 2014;**15**:9. DOI: 10.1186/1467-4866-15-9

[30] Hatcher L, Stepanski EJ. A Step-By-Step Approach to Using the SAS System for Univariate and Multivariate Statistics. Cary: SAS Institute Inc.; 1994. p. 552

[31] Besoain E. Mineralogía de Arcillas de Suelos. San José: IICA; 1985. p. 1205

[32] White GN, Dixon JB. Kaolin-serpentine minerals. In: Dixon JB, Schulze DG, editors. Soil Mineralogy with Environmental Applications. Madison: Soil Science Society of America; 2002. pp. 389-414

[33] Sweegers C, de Coninck HC, Meekes H, Van Enckevort WJP, Hiralal IDK, Rijkeboer A. Morphology, evolution and other characteristics of gibbsite crystals grown from pure and impure aqueous sodium aluminate solutions. Journal of Crystal Growth. 2001;**233**:567-582. DOI: 10.1016/S0022-0248(01)01615-3

[34] Dixon JB. Kaolin and serpentine group minerals. In: Dixon JB, Weed SB, editors. Minerals in Soil Environments. Madison: Soil Science Society of America; 2002. pp. 467-525

[35] Rosenqvist J, Casey WH. The flux of oxygen from the basal surface of gibbsite [α-$Al(OH)_3$] at equilibrium. Geochimica et Cosmochimica Acta. 2004;**68**:3547-3555. DOI: 10.1016/j.gca.2004.02.022

[36] Stumm W, Morgan J. Aquatic Chemistry: An Introduction Emphasizing Chemical Equilibria in Natural Waters. 2nd ed. New York: Wiley-Interscience; 1981. p. 780

[37] Adekola F, Fedoroff M, Geckeis H, Kupcik T, Lefevre G, Lützenkirchen J, et al. Characterization of acid-base properties of two gibbsite samples in the context of literature results. Journal of Colloid and Interface Science. 2011;**354**:306-317. DOI: 10.1016/j.jcis.2010.10.014

[38] Helyar KR, Munns DN, Burau RG. Adsorption of phosphate by gibbsite. I. Effects of neutral chloride salts of calcium, magnesium, sodium, and potassium. Journal of Soil Science. 1976;**27**:307-314. DOI: 10.1111/j.1365-2389.1976.tb02001.x

[39] Brady PV, Cygan RT, Nagy KL. Molecular controls on kaolinite surface charge. Journal of Colloid and Interface Science. 1996;**183**:356-364. DOI: 10.1006/jcis.1996.0557

[40] Lützenkirchen J. Ionic strength effects on cation sorption to oxides: Macroscopic observations and their significance in microscopic interpretation. Journal of Colloid and Interface Science. 1997;**195**:149-155. DOI: 10.1006/jcis.1997.5160

[41] Muljadi D, Posner AM, Quirk JP. Mechanism of phosphate adsorption by kaolinite gibbsite and pseudoboehmite. I. Isotherms and effect of pH on adsorption. Journal of Soil Science. 1966;**17**:212-228. DOI: 10.1111/j.1365-2389.1966.tb01467.x

[42] Chen YSR, Butler JN, Stumm W. Adsorption of phosphate on alumina and kaolinite from dilute aqueous-solutions. Journal of Colloid and Interface Science. 1973;**43**:421-436. DOI: 10.1016/0021-9797(73)90388-3

[43] Manning BA, Goldberg S. Modeling arsenate competitive adsorption on kaolinite, montmorillonite and illite. Clays and Clay Minerals. 1996;**44**:609-623. DOI: 10.1346/Ccmn.1996.0440504

[44] Gérard F. Clay minerals, iron/aluminum oxides, and their contribution to phosphate sorption in soils—A myth revisited. Geoderma. 2016;**262**:213-226. DOI: 10.1016/j.geoderma.2015.08.036

[45] He LM, Zelazny LW, Baligar VC, Ritchey KD, Martens DC. Ionic strength effects on sulfate and phosphate adsorption on γ-alumina and kaolinite: Triple-layer model. Soil Science Society

of America Journal. 1997;**61**:784-793. DOI: 10.2136/sssaj1997.0361599500610 0030011x

[46] Bolland MDA, Posner AM, Quirk JP. pH-independent and pH-dependent surface-charges on kaolinite. Clays and Clay Minerals. 1980;**28**:412-418. DOI: 10.1346/Ccmn.1980.0280602

[47] Geckeis H, Lützenkirchen J, Polly R, Rabung T, Schmidt M. Mineral-water interface reactions of actinides. Chemical Reviews. 2013;**113**:1016-1062. DOI: 10.1021/cr300370h

[48] Van Raij B, Peech M. Electrochemical properties of some Oxisols and Alfisols of tropics. Soil Science Society of America Proceedings. 1972;**36**:587-593. DOI: 10.2136/sssaj1972. 03615995003600040027x

[49] Dzombak DA, Morel FMM. Surface Complexation Modeling: Hydrous Ferric Oxide. Hoboken: John Wiley & Sons Inc.; 1990. p. 393

[50] Riese AC. Adsorption of radium and thorium onto quartz and kaolinite: A comparison of solution/surface equilibria models [thesis]. Bolder: Colorado School of Mines; 1982

Chapter 4

Photocatalytic Adsorbents Nanoparticles

Gustavo Lopes Colpani, Adrieli Teresinha Odorcik Dal'Toé, Micheli Zanetti, Rubieli Carla Frezza Zeferino, Luciano Luiz Silva, Josiane Maria Muneron de Mello and Márcio Antônio Fiori

Abstract

Photocatalysis and high adsorption coupling in a same nanoparticle have been emerged as a prominent class of cost-effective materials to degrade recalcitrant contaminants in wastewater. α-Hematite, metal-organic frameworks and TiO_2 nanocomposites have been investigated due to their features that overcome the other conventional photocatalysts and adsorbents to remove contaminants in aqueous medium. Several methods are applied to synthesize these nanostructures with different properties and physicochemical features and a brief review is shown to these well-established techniques to provide an understanding for the construction and application of these advanced materials.

Keywords: photocatalytic adsorbents, α-hematite, metal-organic frameworks, TiO_2, nanoparticles

1. Introduction

Most conventional technologies applied to the removal of contaminants dispersed in the aqueous effluent employ the adsorption techniques. However, a class of contaminants, not yet regulated in relation to disposal parameters and/or presence in the environment by the government agencies, may have potentially harmful effects on aquatic biota and humans. However, the adsorption techniques need to be investigated in the removing processes of an important class of contaminants that not yet was regulated in relation to discard in the environment by the government agencies, but that have potentially harmful effects on aquatic biota and humans. These emerging contaminants may be synthetic or natural and their most prominent compounds are the drugs, the personal care products, hormones, flame retardants, perfluoroalkylated substances and pesticides [1, 2]. These compounds are called emerging contaminants and their nomenclature originates from the low concentrations detected in surface waters ($\mu g{\cdot}l^{-1}$ or $ng{\cdot}l^{-1}$) [3, 4]. The incidences in very small quantities could only be identified in the last 20 years, due to the appearance of new analytical equipment [5].

IntechOpen

Compounds of difficult degradation by conventional sewage processes can affect the lives of millions of people due to contamination of water resources [6]. The incidence of recalcitrant molecules introduced into water resources is a serious problem of toxicity and deserves the development and research of specific affluent and effluent treatment systems, with greater process efficiency, greater safety and lower costs [7, 8]. Physicochemical and biological methods currently employed are capable of degrading some of these compounds, but they have a high-cost application, which makes it difficult to implement in regions with limited financial resources [9]. Furthermore, these methods can generate complex residues for treatment or storage, such as sludge and toxic or recalcitrant compounds [10].

Thus, alternative methods have been researched in an attempt to eliminate these contaminants effectively, as the adsorption techniques. The adsorption methods have some important vantages: can be operated with low energy consumer and with low operational cost, and cannot generate secondary compounds [11, 12]. The adsorbents applied in the adsorption techniques must have a high surface area, an adequate pore density and surface charge capable of interacting with specific compounds to be removed from the aqueous solution, which makes assure a high removal capacity [13, 14]. These adsorbents are produced from a variety of sources, such as clays, zeolites, silica and carbonaceous materials, which can be applied directly or after chemically or physically activation [15].

However, these processes need be improved because additional costs are still necessary for the waste disposals or for the regeneration of the adsorbents [16]. But recent combinations with other technologies have been reported to improve the removal of the organic pollutants from the water by nanoparticles adsorbents. One of the most attractive methods with nanoparticles is the combination of the adsorption processes and the advanced oxidation processes (AOPs) to ensure high efficient for the capture and for the degradation of the contaminants [17]. Advanced oxidative processes are characterized mainly by the production of •OH radicals in the reaction media or on the catalyst surface of the nanoparticles, which are oxidative species with high redox potential (E^0 = 2.80 V). Due to this feature, these radicals have capacity to degrade a great variety of organic compounds, transforming them into smaller molecules or reducing them completely in carbon dioxide (CO_2) and water (H_2O), mineralizing them [18, 19].

Among the advanced oxidation process, heterogeneous photocatalysis is widely researched with the nanoparticles adsorbents. In this process, very reactive oxygen species (•OH, $^1O_2$ e •O_2^-) are generated from the use of semiconductor catalysts nanostructures induced by ultraviolet or visible radiation [20]. The semiconductors capacity to absorb energy and degrade molecules is due to the energy bands created by atomic orbitals arranged when a new compound is formed on the surfaces of the nanoparticles. These regions are denominated conduction band (CB) and valence band (VB) [21]. In the semiconductors, there will be a region without available energy levels, which are described as band gap [22]. In these cases, only the nanoparticles will be a catalyst if an amount of energy equal or greater than the band gap is provided, which will allow electrons transfer from valence band to conduction band, as depicted in **Figure 1**. The electron (e^-) in the CB can interact with oxygen molecules and form reactive groups, such as the hydroxyl (•OH) and superoxide (•O_2^-) radicals and singlet oxygen ($^1O_2$) [23–25]. On the other hand, the vacancies (h^+) created due to the electrons transfer are oxidants with potential to convert water molecules in •OH [26].

Therefore, a nanoparticle that can combine the adsorption affinity with photocatalysis will give rise to novel materials, which would join the advantages of both techniques to degrade organic pollutants, instead of only change the phase of these contaminants [27, 28]. The major photocatalytic adsorbents researched are α-hematite (Fe_2O_3), meta-organic frameworks (MOFs) and TiO_2 nanocomposites, which are described further.

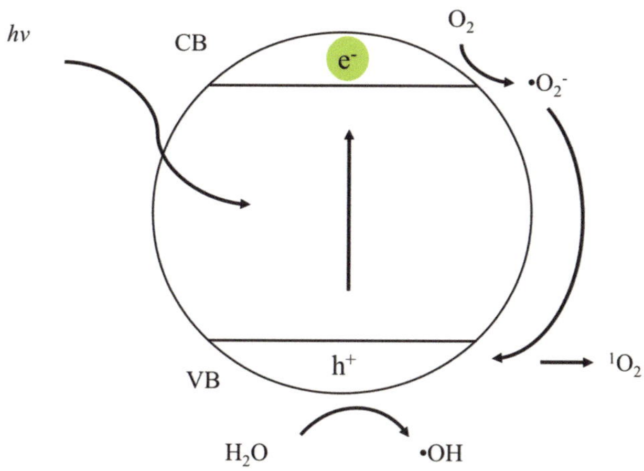

Figure 1.
Semiconductors photocatalytic mechanism.

2. α-Hematite (Fe_2O_3)

Hematite is the most thermodynamically stable phase of iron oxide and is formed by a hexagonal crystalline structure. It is used in catalysis processes and as pigments, in the lithium ion batteries and for the water treatment as an adsorbent [29, 30]. The strong adsorption capacity of organic and inorganic species by the hematite is related to its intrinsic chemical and electronic properties [31, 32]. The pH ZPC of the hematite is approximately 8.0, which implies that in aqueous media with pH below 8.0, the surface is positively charged and the interaction between anionic molecules is stronger [30, 31].

This semiconductor is also employed in photocatalysis due to its reduced band gap energy (E^0 = 2.2 eV), which allows the visible light absorption and with an energy level higher than water oxidation potential. Moreover, the α-Fe_2O_3 phase is a material with low cost and environment friendly and has good chemical and corrosion stability [33]. It is found that because of these characteristics, the hematite nanoparticles are potential candidates to good adsorbents for the removal of contaminants from aqueous systems.

Many methods have been researched and applied to produce hematite nanoparticles, but the most common techniques are the thermal decomposition, sol-gel, solvothermal and hydrothermal [34–36]. These different methodologies are used to control parameters like size, shape and crystal structure of the particles produced, with the aim to ensure interesting adsorption capacity and photocatalysis [37].

Thermal decomposition usually applies inorganic salts, such as nitrates, oxalates and chlorides, with low thermal stability to produce metal oxide nanoparticles in air atmosphere [38]. However, this process can be accomplished by calcination of magnetite (Fe_3O_4) or maghemite (γ-Fe_2O_3) at 500°C [39]. Al-Gaashani et al. [40] synthesized hematite by thermal decomposition of iron (III) nitrate 9-hydrate at different temperatures (300, 400, 500 and 600°C) under ambient conditions within a short time (20 min). The nanostructures showed crystallite sizes in a range to 10–30 nm with high surface area and optical band gap near 3.0 eV. The iron oxide α-Fe_2O_3 nanoparticles with different crystalline structure were prepared by [41], which calcinated the iron precursor ferric ammonium citrate at different temperatures (300, 400 and 550°C) under air atmosphere. The mean particle sizes of the samples were near 20 nm with excellent dye photocalytic degradation under visible light after 10 h. Yang et al. [42] applied direct thermal decomposition of goethite

(α-FeOOH, calcinated at 300°C, to produce hematite with a high adsorption capacity for arsenic (14.5 mg•g^{-1}).

In the sol-gel method, a metal salt, usually chloride and nitrate, is dispersed in alcohol solution to form the "sol" phase by hydrolysis, condensation and polymerization. The "sol" phase is heated to evaporate the solvent and this mixture is converted to "gel" phase, which is composed for regular nanoparticles [43, 44]. Xiao et al. [45] evaluated the adsorption of chromium (VI) on hematite nanoparticles produced by sol-gel technique. The precursors used to form hematite were $FeCl_3$•$6H_2O$ and $(NH_4)_2CO_3$ dissolved in water and ethanol. The crystalline size of α-Fe_2O_3 nanoparticles samples showed values between 15 and 30 nm, with a removal efficiency of chromium near 50% to smaller particles. Sol-gel method was applied by [46] to form hematite nanoparticles from $Fe(NO_3)_3$•$9H_2O$ dispersed in ethanol. The particles produced by this method possess sizes below 10 nm and excellent optical properties, with a band gap energy near 2 eV. Raja et al. [47] employed ferric nitrate dissolved in ethylene glycol to produce hexagonal nanoparticles with interesting structural, optical and magnetic properties. The photocatalytic properties of hematite produce by sol-gel technique were evaluated by [44], that used $Fe(NO_3)_3$•$9H_2O$ dispersed in ethylene glycol to form nanoparticles with a crystalline size smaller than 31 nm and specific surface area between 37 and 57 m^2•g^{-1}. The best photocatalytic activity for H_2 production was obtained on α-Fe_2O_3 calcined at 500°C, with an average evolution rate of 0.015 cm^3•h^{-1}•(mg catalyst)$^{-1}$.

Solvothermal methodology is a versatile process to synthesize uniform-sized nanoparticles, which are formed from iron salts dissolved in non-aqueous solvents, usually alcohol. This mixture often is transferred to autoclave to achieve reaction temperature near 180°C [38, 43]. Preparation of hematite nanoparticles through alcoholysis of ferric chloride under solvothermal condition has been carried out by [48]. $FeCl_3$•$6H_2O$ was dissolved in ethanol and transferred into Teflon autoclave to achieve 180°C. This synthetic method applied created core/shell structures formed from α-Fe_2O_3 nanoparticles, which were used as catalysts in the oxidation of benzyl alcohol to benzaldehyde with 42% conversion and 95% selectivity. Sun et al. [49] developed a combined solvothermal/microwave method to prepare α-Fe_2O_3 nanosheet-assembled hierarchical hollow mesoporous microspheres from chloride ferric precursors dissolved in ethylene glycol. The reaction occurred into autoclave at 180°C for 180 min and the nanoparticles formed were applied to salicylic acid degradation, with removal efficiencies near 50%.

Hydrothermal is the most applied process in the synthesis of hematite due to low cost and energy consumption, high purity and short time to preparation. In this method, the iron (III) chloride or nitrate solutions are dispersed in aqueous media that contains ammonia to adjust the pH value above 9.0. Later, the solution is heated at high temperatures ($\approx$ 150°C) in an autoclave to decompose the precursors and combine their ions to form new compounds with high homogeneity [50, 51]. Hematite nanoparticles were synthesized by [52] by a simple hydrothermal synthesis method using only $Fe(NO_3)_3$•$9H_2O$ and NH_3•H_2O as raw materials into a Teflon autoclave at different temperatures (80, 100, 120 and 150°C) for 10 h. This research demonstrated that nanoparticles with diameter of 30–100 nm only were produced at temperatures above 120°C in a reaction time within 5 h, occurring an increase in the particles size when longer times were evaluated. The methylene blue adsorption on surface of nanostructured t-ZrO2-modified α-Fe_2O_3 composite synthesized in a mesoporous structure by a hydrothermal route was studied by [53]. The composite was prepared from $FeCl_3$•$6H_2O$ and $ZrOCl_2$•$8H_2O$ dissolved in an aqueous media with cetyl trimethyl ammonium bromide (CTAB). Thereafter, the reaction mixture was transferred to a Teflon autoclave and heated at 180°C for 48 h. The nanoparticles synthesized showed excellent adsorption capacity with an efficient upper 95% in the removal of chromium, cobalt, nickel, cadmium, lead, copper and mercury. Dong et al. [54] obtained grain-like hematite by hydrothermal approach using $FeCl_3{\cdot}6H_2O$ and urea dissolved in a water-ethanol solution heated at 180°C for 8 h in a Teflon autoclave. The researchers determined that in a volume ratio of ethanol/water equal to 1:1, the grain-like nanoparticle formation is favored, with no regular structures, when only water or alcohol were used. The photocatalyst response achieved approximately 99% of dye removal after 24 h under visible light.

The researches above-mentioned investigated the adsorption capacity and photocatalysis separately, however the hematite is interesting to organic and inorganic compound removal in water due to the join between both systems, since this photocatalytic adsorbent has a system that captures and degrades the contaminants efficiently. The conjugated response of hematite was evaluated by [55] that synthesized nanoparticles with 75 nm and a specific surface area near 25 $m^2{\cdot}g^{-1}$, which showed a high adsorption capacity (84 $mg{\cdot}g^{-1}$) and limited photocatalytic degradation (≈ 20%).

Cheng et al. [56] also analyze the conjugated adsorption and photocatalytic removal of organic dye rhodamine B for cauliflower-like α-Fe_2O_3 microstructures constructed by nanoparticle-based buds. The hematite nanoparticles were synthesized from aqueous solution containing ammonia and polyvinylpyrrolidone (PVP) that was mixed with Fe(acetylacetonate)$_3$ dispersed in a toluene solution and transferred to autoclave at 150°C for 24 h. The structure formed with high surface area (31.57 $m^2{\cdot}g^{-1}$) were shown to exhibit adsorption efficient near 80% of organic dye in wastewater and structurally enhanced visible light photocatalytic activity, with a degradation above 90% in the presence of H_2O_2.

Liu et al. [57] prepared α-Fe_2O_3 hollow spheres with novel multiple porous shells by solvothermal treatment of $FeCl_3{\cdot}6H_2O$ dissolved in ethylene glycol and CH_3COONa. Carbon spheres were added into the mixture and the resultant suspension was transferred to an autoclave at 200°C for 10 h. The carbon was removed by calcination at 500°C during 3 h. The nanostructures formed are composed of particles from 20 to 40 nm in size that reduced the dye concentration below 90% due to the synergistic effect of adsorption and photocatalysis.

Hematite nanoparticles were synthesized by [58] applying hydrothermal method to react $FeCl_3{\cdot}6H_2O$ dissolved in aqueous solution containing NH_4OH and pectin into a Teflon autoclave. The size of the nanoparticles was achieved as 42 nm when pectin was added probably due to less agglomeration. The experimental data demonstrated that dye removal reached values upper 80% when only adsorption was applied, as well as when the photocatalysis is incorporated.

3. Metal-organic frameworks (MOFs)

Metal-organic frameworks (MOFs) are a class of functional materials synthesized by the assembly of the metal ions/clusters and organic linkers like cyano and pyridyl, carboxylates, phosphonates and crown ethers, which are connected to metal ions/clusters through coordination bonds of moderate strength [59, 60]. The MOFs have been studied since 1990s with more than 20,000 structures synthesized and evaluated for applications such as adsorption, catalysis, drug delivery, sensing, separation, gas storage, bioimaging and so on [61–64]. These structures have shown huge potential in these areas due to distinctive features, such as high porosity and surface area, chemically adjustable pore, uniform structures, tunable surface properties (functional groups), good thermal stability, unsaturated metal centers and even the catalytically active organic linkers [65–68]. The most typical metal-organic frameworks are MIL (Materials of Institute Lavoisier), based on lanthanides or transition metals; UiO (University of Oslo), built up with Zr; MOF-5, composed of Zn; and Cu-BTC, based on Cu [69].

These structures have been applied to adsorption of organic and inorganic contaminants in water and wastewater with better efficiencies when compared with conventional adsorbents, probably due to large surface area (1000–10,000 $m^2 \cdot g^{-1}$); presence of central metal ions, open metal sites, coordinatively unsaturated sites and functional groups on the organic linkers; and easily tunable structure and flexible framework, since it is possible to modify the pores surface [66, 70–72]. The adsorption mechanisms for removal of organic pollutants by MOFs in water mainly include electrostatic, hydrophobic, acid-base π-π interactions and hydrogen bonding [73–75].

MOFs photocatalytic properties are exploited due to high capacity of the organic linkers in absorbing photons, such as antennas to harvest light, which transfer the energy to the metal sites by transition from ligand to metal cluster charge under UV or visible light radiation [76, 77]. Therefore, these structures can act as semiconductor, since MOFs contain conduction and valence bands, with band gaps in the range of 1.0–5.5 eV, or the band gaps are related to the energy levels of the highest occupied molecular orbitals (HOMOs) and lowest unoccupied molecular orbitals (LUMOs) of the linker molecules, which leads to the formation of charge carriers (e^-/h^+) that can subsequently be transferred to the surface [60, 68]. Furthermore, the easily tailorable physical and chemical functions, together with the large surface area and permanent pores/channels to potentially anchor/encapsulate photosensitizers and catalytic moieties, make MOFs potential candidates for application in photocatalytic processes [78].

The photocatalytic and adsorption features of the metal-organic frameworks are directly related to synthetic routes and parameters such as temperature, reagent concentration, solvent, pH and pressure [79]. Many techniques have been studied to produce distinct structures with specific properties, such as hydro/solvothermal, electrochemical, mechanochemical and solvothermal [80–82].

Electrochemical method to synthesize MOFs is based on the dissolution of a anodic metal that supply metal ions into a reaction medium that contains the organic linkers and electrolytes, avoiding the use of metal salts. This reaction between metal ions and organic linkers produces structures with high purity, due to the absence of nitrate, perchlorate or chloride presents in metal salts, in a short time when compared to other synthetic methods [83–85]. The electrochemical approach was applied by [86] to produce $Cu_3(BTC)_2$ (HKUST-1) for CO_2 and CH_4 adsorption and separation. Two copper electrodes were used into a solution that contained benzene-1,3,5-tricarboxylate (H_3BTC) and tetrabutylammonium tetrafluoroborate (TBAFB) as electrolytes to form the MOF. The characterization confirmed the high textural properties, high crystallinity and good thermal stability of the structure, which showed satisfactory CO_2 adsorption. A microseparator device containing a metal-organic framework synthesized by electrochemical method was produced by [87]. HKUST-1 was prepared by copper electrodes, benzene-1,3,5-tricarboxylate (BTC) linker and methyl-tributyl-ammonium methyl sulfate (MTMS) electrolyte and was applied to the separation of methanol and n-hexane. The breakthrough curves analysis demonstrated that up to 400 $mg \cdot g^{-1}$ of methanol can be adsorbed to the electrochemically synthesized Cu-BTC coating. Yang et al. [88] synthesized MOF-5 using zinc anodes, terephthalic acid (H_2BDC) linker, zinc nitrate hexahydrate [$Zn(NO_3)_2 \cdot 6H_2O$] electrolyte and 1-butyl-3-methylimidazolium chloride ionic liquid, which was employed to a template to induce the porous structure. This structure was mixed with BiOBr by ultrasound treatment to create a composite that was evaluated for methyl orange dye degradation by photocatalysis under simulated solar light irradiation. The results showed that this new composite can achieve degradation more than 90% that occurs mainly because of the H^+ and $\cdot O_2^-$ active species.

Mechanochemical method is a solvent free methodology to produce MOFs that employ a direct mechanical grinding of the metal salts and linker precursors either in a mortar or a ball mill [89]. This technique can occur at room temperature in short reaction times [90]. The compound MOF-14 [$Cu_3(BTB)_2$] was synthesized by [91] using ball milling to copper acetate monohydrate (metal salt) and H_3BTB (organic linker) mixture, which were placed in a ball mill with three balls and grinded together for 10 min and later activated by a single post-synthetic washing step with ethanol. This structures showed high micropore volume and a specific surface area near 1200 $m^2{\cdot}g^{-1}$. Framework $Zn_2(oba)_2(4\text{-bpdb}){\cdot}(DMF)_x$ (TMU-4) and $Zn_2(oba)_2(4\text{-bpdh}){\cdot}(DMF)_y$ (TMU-5) were prepared by mechanochemical by [92] from a mixture of 4,4′-oxybisbenzoic acid (H_2oba) and 1,4-bis(4-pyridyl)-2,3-diaza-1,3-butadiene (4-bpdb) or 2,5-bis(4-pyridyl)-3,4-diaza-2,4-hexadiene (4-bpdh), respectively, for 15 min. The structures obtained possess different structural topologies, metal-ligand connectivities and therefore different pore sizes. Furthermore, the pore surface in both networks were capable to capture CO_2, achieving 60 $cm^3{\cdot}g^1$. Chen et al. [93] synthesized $[In_2(OH)_2(BPTC)]{\cdot}6H_2O$ (InOF-1) by mechanochemical route from indium acetate hexahydrate $In(OAc)_3{\cdot}6H_2O$ and organic linker 3,3′,5,5′-biphenyltetracarboxylic acid (H_4bptc), which were filled in a stainless steel milling jar for different times (10–60 min). InOF-1 showed moderate adsorption capacity for CO_2 and high for CO_2/CH_4 and CO_2/N_2 adsorption selectivities.

In the sonochemical technique, the MOF synthesis occurs due to extremely high temperature ($\approx$ 4000 K) and pressure ($\approx$ 1000 atm) in microenvironment formed by acoustic cavitation generated by ultrasound, which starts the chemical bonds breakage of the elements in the solution and allows interaction between metal salts and organic linkers [94, 95]. This is an environment friendly method to produce homogeneous nucleation centers in a short time and with low energy consumption [79]. Zn(II)-based metal-organic framework $[Zn(TDC)(4\text{-BPMH})]_n{\cdot}n(H_2O)$ was produced by [96] from 2,5-thiophene dicarboxylic acid (TDC) linker, N,N-bis-pyridin-4-ylmethylene-hydrazine (4-BPMH) as pillar spacer and zinc nitrate hexahydrate $[Zn(NO_3)_2{\cdot}6H_2O]$ as metal precursor. The MOFs nanoparticles were sonochemically synthesized under atmospheric pressure using an ultrasonic bath with different irradiation powers, irradiation time, precursor concentrations and temperatures. Adsorption capacities of 2,4-dichlorophenol and amoxicillin were evaluated in aqueous media and their related results demonstrated an efficiency nearly 90% for both cases. Sonochemical method was employed by [97] to synthesize $[Cd(oba)(4\text{-bpdh})]_n{\cdot}1DMF$ (TMU-7) using the ligand 4,4-oxybisbenzoic acid (H_2oba), the N-donor ligand 2,5-bis(4-pyridyl)-3,4-diaza-2,4-hexadiene (4-bpdh) and the metal precursor zinc acetate tetrahydrate $[Zn(OAc)_2{\cdot}4H_2O]$. Synthesis of TMU-7 was carried out in an ultrasonic bath at ambient temperature and atmospheric pressure with different reaction times and concentrations of metal and ligands to evaluate the morphology. The results showed that the Congo red dye was efficiently removed when this nanostructures were used, achieving an adsorption approximately equal 97%. Abdollahi et al. [98] produced $Zn_4(oba)_3(DMF)_2$ from $Zn(NO_3)_2{\cdot}6H_2O$ and H_2oba dispersed in DMF. MOFs syntheses were carried out in an ultrasonic bath at ambient temperature and atmospheric pressure for different reaction times and concentrations of initial precursors. In comparison of samples synthesized by solvothermal and sonochemical methods, the nanostructures synthesized by ultrasound have been more efficient to remove Congo red and Sudan red dyes, with an adsorption efficiency approximately 53 and 87%, respectively.

Hydro/solvothermal synthesis is the most classical route to produce metal-organic frameworks nanoparticles. This method involves the heterogeneous reaction between organic linkers and metal salts that are dissolved in water or organic solvents (e.g. alcohols and pyridine) under moderate to high temperatures and pressures [84]. This reaction occurs in sealed reactor vessels (autoclave) under conditions above the boiling point of the solvent and usually takes place over days or hours [99, 100]. The frameworks synthesized by hydro/solvothermal synthesis are nanoparticles with high crystallinity and good size control, which are insoluble in the solvent [70]. Mn (II) ions based metal-organic framework was synthesized by solvothermal method described in [101]. A solution of 1,3,5-tris(4-carboxyphenyl) benzene acid (H_3BTB) as linker, manganese (II) acetate [$Mn(OAc)_2$] as metal salt, imidazole and d L-N-tert-butoxycarbonyl-2-(imidazole)-1-pyrrolidine (L-BCIP) as chiral adduct were sealed in a Teflon stainless steel vessel at 110°C for 3 days. The adsorption rate for methylene blue was nearly 90% after 120 min, when the MOF was used. The zinc-based MOF ($[Zn_5(FODC)_2(OCH_2CH_2O)_3(H_2O)]\cdot(sol)_n$) was synthesized by solvothermal method [102] and evaluated for adsorption and selective separation of methylene blue (MB), crystal violet (CV) and rhodamine B (RhB). Fluorenone-2,7-dicarboxylate (H_2FODC) ligand and $Zn(OAc)_2\cdot 4H_2O$ were mixed and transferred to stainless steel reactor with a Teflon stainless steel autoclave, sealed and heated up to 110°C for 3 days. The structures produced showed high affinity to cationic dyes, with removal rates of MB, CV and RhB on the Zn-MOF equal to 98.44, 90.77 and 41.99%, respectively. The photocatalytic response of $[Ni(azp)(ppa)(H_2O)_2]_n$ metal-organic framework was studied by [103]. This structure was produced from the salt metal nickel nitrate hexahydrate ($[Ni(NO_3)_2]\cdot 6H_2O$), the O-donor ligand 1,4-phenylenedipropionic acid (PPA) and four N,N′-donors ligands [4,4′-azodipyridine (AZP); 4,4′-trimethylenedipyridine (TMDP); 1,2-bis-(4-pyridyl)ethane (BPETHA) and 4,4′-bipyridine (BPY)]. The solutions were individually placed in a Teflon-lined stainless steel heated at 80°C for 10 h and then continuously heated at 120°C for 24 h. The new structures formed displayed diverse structural architectures due to a variety of the length and flexibility of N,N′-donors ligands and the varied coordination modes of PPA. The band gaps of the MOFs were between 3.3 and 3.7 eV, with a photocatalytic degradation of MB roughly in the range of 65–97%. Gao et al. [104] evaluated the photocatalytic degradation of Acid Orange 7 dye in aqueous solution over MIL-53(Fe) under visible LED light irradiation and in the presence of persulfate oxidant. The framework was synthesized by solvothermal technique from a mixture of $FeCl_3\cdot 6H_2O$, 1,4-benzenedicarboxylic acid (H_2BDC) organic linker and DMF. The reactant mixture was transferred into a Teflon-lined stainless steel autoclave and heated at 150°C for 15 h. MIL-53(Fe) band gap was determined equal to 2.62 eV and the photocatalytic degradation response demonstrated that this structure can degrade Acid Orange 7 via the direct hole oxidation pathway under visible light, achieving a removal efficiency upper 90% after 90 min.

Nevertheless, such as above mentioned, for the removal and degradation processes is interesting the evaluation of adsorption and photocatalytic effects coupled. In this context, Gao et al. [105] evaluated the adsorption and visible light photodegradation of aqueous clofibric acid (CA) and carbamazepine (CBZ) by MIL-53(Fe) metal-organic framework prepared by solvothermal method. MOF synthesis was performed from a mixture of $FeCl_3\cdot 6H_2O$, terephthalic acid and DMF that was introduced in a Teflon-lined steel autoclave and maintained at 120°C for 3 days. The adsorption and photocatalytic experiments evaluated the pH effect and the related results suggested that the adsorption of CA and CBZ were mainly ascribed to electrostatic interactions and π–π interactions, respectively. In pH equal 3.0, the maximum adsorption capacity of CA and CBZ on MIL-53(Fe) were 0.80 and 0.57 $mmol\cdot g^{-1}$, respectively, and the photodegradation efficiency for CA and CBZ was greater than 90%, when H_2O_2 was used as oxidant.

Araya et al. [106] synthesized the FeBTC MOF modified with Amberlite IRA-200 resin to yield a novel heterogeneous photocatalyst, A@FeBTC, to degrade

Rhodamine B dye. The iron-based framework was synthesized using the hydrothermal method from H_3BTC, iron powder, hydrofluoric acid, nitric acid and water. The reaction mixture was transferred to a Teflon-lined pressure vessel and maintained at 160°C for 12 h. A@FeBTC catalyst was prepared by grinding and stirring the powdered resin in an aqueous suspension of FeBTC for 24 h, with different resin/FeBTC mass ratio. All samples (with or without Amberlit resin) were evaluated to dye removal and showed adsorption efficiency greater than 20% and photodegradation responses upper 95% for A@FeBTC and FeBTC.

Simultaneously efficient adsorption and photocatalytic degradation of tetracycline by Fe-MIL-101, Fe-MIL-100 and Fe-MIL-53 was studied by [107]. MOFs were synthesized by a hydrothermal method from $FeCl_3{\cdot}6H_2O$ and H_2BDC dissolved in DMF. The solution was sonicated, transferred to a Teflon-lined stainless steel autoclave and maintained at 110°C in an oven for 20 h for Fe-MIL-101 and maintained at 150°C for 12 h for Fe-MIL-53. In the synthesis of Fe-MIL-100, a mixture of $FeCl_3{\cdot}6H_2O$, H_2BDC and hydrofluoric acid was dissolved in DMF and magnetically stirred. The solution was heated at 160°C for 12 h into a Teflon-lined stainless steel autoclave. The effects of adding dosage and initial concentration of tetracycline on degradation efficiency were examined and the results revealed that a Fe-MIL-101 catalyst dosage equal 0.5 $g{\cdot}l^{-1}$ showed the best photocatalytic efficiency. Tetracycline was efficiently adsorbed and degraded by Fe-MIL-101, reaching approximately 55% only by adsorption removal and 96.6% after photocalysis process.

Abdpour et al. [108] evaluated the MIL-100(Fe)@MIL-53(Fe) photocatalytic performance of methyl orange degradation under visible light. Firstly, MIL-100(Fe) was synthesized by solvothermal method from H_2BTC and $FeCl_3{\cdot}6H_2O$ dissolved in distilled water in the presence of HNO_3 and HF. The resulted solution was stirred, transferred to the autoclave and heated at 150°C for 20 h. Then, different amounts of MIL-100(Fe) nanoparticles were dispersed in DMF by ultrasonication. $FeCl_3.6H_2O$ and H_2BDC were added in this solution, which was stirred and placed in the ultrasonic generator probe for 15 min at 50% of the maximum power. The sonochemically synthesized the metal-organic framework was denominated MIL-100(Fe)@MIL-53(Fe). The band gaps energies calculated for the samples were around 2.5 eV. The results showed that all samples reached adsorption efficiencies near 10–15%, whereas the samples that contained 0.03 and 0.04 g of MIL-100(Fe) achieved the highest photocatalytic degradation (≈70%). The reduction of the photogenerated electron-hole pairs recombination probably occurs due to the decrease in photoluminescence intensity because excited electron transfer from the conduction band of MIL-100(Fe) to MIL-53(Fe) and hole transfer from the valance band of MIL-53(Fe) to MIL-100(Fe), which prolong the lifetime of the separated electrons and holes in the MIL-100(Fe)@MIL-53(Fe).

4. α-Fe_2O_3@TiO_2, MOF@TiO_2 and others TiO_2 nanocomposites

TiO_2 application as a photocatalyst is of great interest because this compound is nontoxic, economically viable, chemically inert, photostable to corrosion, besides having high thermal stability and intense photocatalytic activity and oxidation power [22, 109]. The main crystalline structures of titanium dioxide are anatase, rutile and brookite, but the last one is difficult to be synthesized in laboratory [110]. However, not all crystalline structures have the same efficiency in the absorption of light for catalysis, and rutile, although the polymorph thermodynamically more stable has reduced photocatalytic activity in comparison to anatase [111]. This occurs possibly due to the high temperature required for its preparation, resulting in an increase in the particle size, lower electron mobility in relation to anatase and high rate of electron/vacancy recombination, that cause a reduced number of hydroxyl groups on the surface [10].

However, despite all these favorable features, TiO_2 has some limitations that affects its large-scale application in industrial processes, such as the recombination of photogenerated charges due to defects, impurities or other imperfections on crystal surface, which reduce the photocatalytic efficiency. In addition, the band gap energy required for the formation of the electron/vacancy pair is equal to 3.2 eV, which restricts this catalyst to the use of UV-light [112]. Since only 5% of sunlight wavelengths are in the UV region, an alternative energy source is necessary, making the process more expensive when this semiconductor is applied [23]. Furthermore, the reactive oxygen species (ROS) are formed on the surface of the titanium dioxide, mainly due to the reaction and capture of an electron of the water molecule by the vacancy present on this surface or by the electron donation on the CB. Accordingly, the higher density of radicals is close to the semiconductor surface. Therefore, strategies that improve the incidence of the molecules on the surface of the photocatalyst, like adsorption, would increase the probability of attack and degradation of these molecules by ROS. This can enhance the selectivity and the photocatalytic capacity of this advanced oxidation process.

Therefore, many efforts have been performed to improve the adsorption capacity and photocatalytic efficiency of TiO_2. Structural modifications of TiO_2 have been proposed to improve the photocatalysis, such as metal deposition, doping with nonmetals, functionalization with organic molecules or coupling of other metals with TiO_2 [25, 113]. Different methods have been adopted to synthesize TiO_2 nanoparticles such as sol-gel, solvothermal, hydrothermal, sonochemical and mechanochemical, which were described in the above sections [114–116].

Cheng et al. [117] synthesized a Fe_2O_3@TiO_2 nanocomposite with high adsorption and photocatalytic activity by solvothermal method. Titanium glycolate precursor and $(NH_4)_2Fe(SO_4)_2{\cdot}6H_2O$ were dispersed by vigorous sonication in deionized water. This reaction mixture was autoclaved for 6 h at 180°C in a Teflon-lined autoclave. The combined effect of adsorption and photocatalysis were evaluated by Rhodamine B degradation under visible light. It was observed that the TiO_2@Fe_2O_3 exhibited improved adsorption (RhB removal ≈ 20%) and photocatalytic (RhB degradation ≈ 75%) capability compared to pure TiO_2 and Fe_2O_3 nanoparticles.

Fe_2O_3@TiO_2 nanocomposite was prepared via mechanochemical process by [118]. Commercial P25® and different amounts of magnetic Fe_3O_4 nanoparticles were mixed by a mechanical process and the mixtures were calcined at 450°C for 2 h. XRD characterizations revealed that the samples are formed basically for TiO_2 and α-Fe_2O_3 after calcination, with high purity. MB adsorption efficiency is near 15% for 5%-$Fe_2O_3$2TiO_2 with photocatalytic degradation approximately equal to 65% after 80 min under visible light.

TiO_2@MIL-101 core-shell structure was synthesized by [119] under hydrothermal method and investigated for adsorption and photocatalytic degradation of methyl orange. MIL-101 was synthesized under hydrothermal conditions and added with titanium butoxide into ethanol solution under stirring. Then, ultrapure water was added into the solution that was transferred into a Teflon-lined bomb sealed and heated at 220°C for 3 h to produce TiO_2@MIL-101 core-shell composites. The results revealed that the TiO_2@-MIL-101 core-shell composite possessed excellent adsorption of MO (removal ≈ 70%), probably due to π-π interaction between benzene rings in MO and MIL-101, and also showed inspiring property on the degradation of MO, reaching a removal efficiency equal 97% after 50 min under UV-light radiation.

A photocatalytic adsorbent, N-doped TiO_2 nanoparticles encapsulated in MIL-100(Fe) cages was developed by [120] for the adsorption and photocatalytic degradation, enhances methylene blue (MB) and rhodamine B (RhB) under visible light. MIL-100(Fe) was synthesized under hydrothermal technique from a mixture of Fe powder, H_3BTC, HF, HNO_3 and H_2O that was stirred and transferred into a Teflon-lined stainless steel autoclave at 150°C for 12 h. The samples of MIL-100(Fe) encapsulating N-TiO_2 nanoparticles were prepared by impregnating MIL-100(Fe) in a dilute suspension that contained different concentrations of as-prepared neutral N-TiO2 nanoparticles. This mixture was stirred for 3 h at ambient temperature, dried and calcinated at 150°C for 4 h. All the samples showed adsorption efficiencies greater than 80% under dark conditions, probably related to MOFs pore size (≈1.93 nm) and high surface area (≈1400 $m^2 \cdot g^{-1}$). The photocatalytic activity of 32%N-TiO_2@MIL-100(Fe) sample was the highest of all samples evaluated, with an degradation efficiency equal 99.1% for MB and 93.5% for RhB.

However, to improve the adsorption capacity and to enhance the photocatalytic capacity, researchers have evaluated the inclusion of distinct organic and inorganic compounds, such as cyclodextrins and their derivatives, noble metals (silver), lanthanides (cerium, samarium, lanthanum and neodymium), graphene, zeolites, alumina and silica.

In this context, Dal'Toé et al. [121] investigated the incorporation of plasmonic Ag nanoparticles on the physicochemical and photocatalytic properties of La-doped TiO_2 nanostructure. The nanocrystalline La-doped TiO_2 powder was prepared by an ultrasound-assisted wet impregnation method. $La(NO_3)_3 \cdot 6H_2O$ and commercial TiO_2 P25® were dispersed in distilled water and stirred for 30 min. Then, the solution was ultrasonically processed for 3 h. After the ultrasound processing, the solution was heated to the boiling point and left evaporating for 40 min. The as-obtained paste was dried overnight at 120°C. The resultant solid was ground with a pestle to obtain a fine powder, which was calcined at 500°C for 1 h. Then, La/TiO_2 sample was dispersed into distilled water and aliquots of $AgNO_3$ solution was added according to the desired Ag molar ratio (0.5–5%). The mixture was then photoirradiated under 80 W Hg vapor lamp and dried. The results revealed an increase in the adsorption capacity of the nanoparticles when the Ag molar ratio is between 2 and 4%, with efficiencies more than 65%, which proportionally increased the photocatalytic activity. Thus, the enhancements achieved in the photocatalytic decolorization (>95% in 30 min) of MB by Ag-La/TiO_2 materials are directly related to the increased adsorption capacity.

On the other hand, organic molecules functionalization to enhance the photocatalytic activity of TiO_2 was proposed by [122]. Carboxymethyl-β-cyclodextrin (CMCD) functionalization of TiO_2 doped with lanthanum was evaluated to MB adsorption and degradation under UV-light radiation. TiO_2-La nanoparticles were synthesized as described by [121]. The CMCD was synthesized by the dissolution of β-CD and NaOH aqueous $ClCH_2COOH$ solution, which was maintained at 50°C in a jacketed reactor for 5 h. A white precipitate (CMCD) was obtained by addition of methanol and acetone to the solution. The CMCD functionalized catalysts were prepared TiO_2–La dispersed in distilled water. This solution was added to CMCD along with cyanamide and maintained at 90°C for 4 h. The results showed that adsorption efficiency increase for CMCD@TiO_2-La sample, with a removal near 15% after 60 min under dark conditions. Accordingly, the improvement of the photocatalytic activity achieved for this nanoparticle is also related to the adsorption of the MB by CMCD, although the mass transfer is low due to the reduced concentration of this oligosaccharide at the TiO_2 nanoparticles surface. This increase in degradation efficiency occurs because the CMCD has the function of enhancing the density of the dye at the semiconductor and solution interface, where the ROS are formed. Usually these species return to a thermodynamically stable state without reacting with organic molecules due to their short lifetimes and because the contaminants are dispersed in the solution. Thus, when the dye is adsorbed, the transfer of electrons between the ROS and the contaminant becomes more probable, increasing the efficiency of the process.

5. Conclusions

Photocatalytic adsorbents nanoparticles have a tremendous potential to wastewater treatment for capture and degrade many contaminants. Coupling of adsorption and photocatalysis demonstrated to be an efficient process to overcome of the limitations presented by the separated techniques. α-Hematite and metal-organic frameworks have exhibited excellent adsorption performance for removing contaminants from aqueous solutions and superior photocatalytic capacities to degrade and mineralize many recalcitrant compounds in visible light. Some researches have been developed to couple the TiO_2 properties with MOFs or α-Fe_2O_3, as well as other inorganic or organic compounds, to enhance the photocatalytic activity. As illustrated in this review, a range of photocatalytic adsorbents technologies have been proposed or are under active development for wastewater treatment, but many techniques are still at an experimental or pilot stage.

Acknowledgements

The authors acknowledge the Community University of Chapecó and Region (Unochapecó), for technical and scientific support.

Conflict of interest

There are no conflicts to declare.

Author details

Gustavo Lopes Colpani[1], Adrieli Teresinha Odorcik Dal'Toé[2], Micheli Zanetti[3], Rubieli Carla Frezza Zeferino[3], Luciano Luiz Silva[1], Josiane Maria Muneron de Mello[1,4] and Márcio Antônio Fiori[1,4]*

1 Post-Graduation Program in Management and Innovation Technology (PPGTI)—Unochapecó, Chapecó, Brazil

2 Federal University of Santa Catarina—UFSC, Florianópolis, Brazil

3 Chemical and Food Engineering—Unochapecó, Chapecó, Brazil

4 Post-Graduation Program in Environment Science (PPGCA)—Unochapecó, Chapecó, Brazil

*Address all correspondence to: fiori@unochapeco.edu.br

References

[1] Matamoros V, Rodríguez Y, Albaigés J. A comparative assessment of intensive and extensive wastewater treatment technologies for removing emerging contaminants in small communities. Water Research. 2016;**88**:777-785. DOI: 10.1016/j.watres.2015.10.058

[2] Yuan J, Dyke MI, Huck PM. Identification of critical contaminants in wastewater effluent for managed aquifer recharge. Chemosphere. 2017;**172**:294-301. DOI: 10.1016/j.chemosphere.2016.12.120

[3] Naidu R, Espana VAA, Liu Y, Jit J. Emerging contaminants in the environment: Risk-based analysis for better management. Chemosphere. 2016;**154**:350-357. DOI: 10.1016/j.chemosphere.2016.03.068

[4] Noguera-Oviedo K, Aga DS. Lessons learned from more than two decades of research on emerging contaminants in the environment. Journal of Hazardous Materials. 2016;**316**:242-251. DOI: 10.1016/j.jhazmat.2016.04.058

[5] Petrie B, Barden R, Kasprzyk-Hordern B. A review on emerging contaminants in wastewaters and the environment: Current knowledge, understudied areas and recommendations for future monitoring. Water Research. 2015;**72**: 3-27. DOI: 10.1016/j.watres.2014.08.053

[6] Malato S, Ibáñez PF, Maldonado MI, Blanco J, Gernjak W. Decontamination and disinfection of water by solar photocatalysis: Recent overview and trends. Catalysis Today. 2009;**147**:1-59. DOI: 10.1016/j.cattod.2009.06.018

[7] Kaur A, Umar A, Kansal SK. Heterogeneous photocatalytic studies of analgesic and non-steroidalanti-inflammatory drugs. Applied Catalysis A: General. 2016;**510**:134-155. DOI: 10.1016/j.apcata.2015.11.008

[8] Ribeiro AR, Nunes OC, Pereira MFR, Silva AMT. An overview on the advanced oxidation processes applied for the treatment of water pollutants defined in the recently launched Directive 2013/39/EU. Environmental International. 2015;**75**:33-51. DOI: 10.1016/j.envint.2014.10.027

[9] Teh CM, Mohamed AR. Roles of titanium dioxide and ion-doped titanium dioxide on photocatalytic degradation of organic pollutants (phenolic compounds and dyes) in aqueous solutions: A review. Journal of Alloys and Compounds. 2011;**509**:1648-1660. DOI: 10.1016/j.jallcom.2010.10.181

[10] Zangeneh H, Zinatizadeh AAL, Habibi M, Akia M, Isa MH. Photocatalytic oxidation of organic dyes and pollutants in wastewater using different modified titanium dioxides: A comparative review. Journal of Industrial and Engineering Chemistry. 2015;**26**:1-36. DOI: 10.1016/j.jiec.2014.10.043

[11] Nazari G, Abolghasemi H, Esmaieli M, Pouya ES. Aqueous phase adsorption of cephalexin by walnut shell-based activated carbon: A fixed-bed column study. Applied Surface Science. 2016;**375**:144-153. DOI: 10.1016/j.apsusc.2016.03.096

[12] Song JY, Bhadra BN, Jhung SH. Contribution of H-bond in adsorptive removal of pharmaceutical and personal care products from water using oxidized activated carbon. Microporous and Mesoporous Materials. 2017;**243**:221-228. DOI: 10.1016/j.micromeso.2017.02.024

[13] Darweesh TM, Ahmed MJ. Adsorption of ciprofloxacin and norfloxacin from aqueous solution onto granular activated carbon in fixed bed column. Ecotoxicology and

Environmental Safety. 2017;**138**:139-145. DOI: 10.1016/j.ecoenv.2016.12.032

[14] Fayazi M, Ghanei-Motlagh M, Taher MA. The adsorption of basic dye (alizarin red S) from aqueous solution onto activated carbon/γ-Fe_2O_3 nano-composite: Kinetic and equilibrium studies. Materials Science in Semiconductor Processing. 2015;**40**: 35-43. DOI: 10.1016/j.mssp.2015.06.044

[15] Kim Y, Bae J, Park J, Suh J, Lee S, Park H, et al. Removal of 12 selected pharmaceuticals by granular mesoporous silica SBA-15 in aqueous phase. Chemical Engineering Journal. 2014;**256**:475-485. DOI: 10.1016/j.cej.2014.06.100

[16] Mehrjouei M, Müller S, Möller D. A review on photocatalytic ozonation used for the treatment of water and wastewater. Chemical Engineering Journal. 2015;**263**:209-219. DOI: 10.1016/j.cej.2014.10.112

[17] Kanakarajua D, Kockler J, Motti CA, Glass BD, Oelgemöller M. Titanium dioxide/zeolite integrated photocatalytic adsorbents for the degradation of amoxicillin. Applied Catalysis B: Environmental. 2015;**166-167**:45-55. DOI: 10.1016/j.apcatb.2014.11.001

[18] Cheng M, Zeng G, Huang D, Lai C, Xu P, Zhang C, et al. Hydroxyl radicals based advanced oxidation processes (AOPs) for remediation of soils contaminated with organic compounds: A review. Chemical Engineering Journal. 2016;**284**:582-598. DOI: 10.1016/j.cej.2015.09.001

[19] Wng N, Zheng T, Zhang G, Wang P. A review on Fenton-like processes for organic wastewater treatment. Journal of Environmental Chemical Engineering. 2016;**4**:762-787. DOI: 10.1016/j.jece.2015.12.016

[20] Reddy PVL, Kim KH. A review of photochemical approaches for the treatment of a wide range of pesticides. Journal of Hazardous Materials. 2015;**285**:325-335. DOI: 10.1016/j.jhazmat.2014.11.036

[21] Coronado JM, Fresno F, Hernández-Alonso MD, Portela R. Design of advanced photocatalytic materials for energy and environmental applications. In: Green Energy and Technology. 1st ed. London: Springer-Verlag; 2013. 351 p. DOI: 10.1007/978-1-4471-5061-9_1

[22] GAYA UI. Heterogeneous Photocatalysis Using Inorganic Semiconductor Solids. 1st ed. Dordrecht: Springer Science + Business Media; 2014. 222 p. DOI: 10.1007/978-94-007-7775-0

[23] Zhao C, Pelaez M, Dionysiou DD, Pillai SC, Byrne JA, O'shea KE. UV and visible light activated TiO_2 photocatalysis of 6-hydroxymethyluracil, a model compound for the potent cyanotoxin cylindrospermopsin. Catalysis Today. 2014;**224**:70-76. DOI: 10.1016/j.cattod.2013.09.042

[24] Spasiano D, Marotta R, Malato S, Fernandez-Ibañez P, Somma I. Solar photocatalysis: Materials, reactors, some commercial, and pre-industrialized applications. A comprehensive approach. Applied Catalysis B: Environmental. 2015;**170-171**:90-123. DOI: 10.1016/j.apcatb.2014.12.050

[25] Wang W, Huang G, Yu JC, Wong PK. Advances in photocatalytic disinfection of bacteria: Development of photocatalysts and mechanisms. Journal of Environmental Science. 2015;**34**: 232-247. DOI: 10.1016/j.jes.2015.05.003

[26] Marschall R, Wang L. Non-metal doping of transition metal oxides for visible-light photocatalysis. Catalysis Today. 2014;**225**:111-135. DOI: 10.1016/j.cattod.2013.10.088

[27] Zhang Q, Lu Z, Jin S, Zheng Y, Ye T, Yang D, et al. TiO_2 nanotube-carbon macroscopic monoliths with multimodal porosity as efficient recyclable photocatalytic adsorbents for water purification. Materials Chemistry and Physics. 2016;**173**: 452-459. DOI: 10.1016/j.matchemphys. 2016.02.037

[28] Jing HY, Wen T, Fan CM, Gao GQ, Zhong SL, Xu AW. Efficient adsorption/photodegradation of organic pollutants from aqueous systems using Cu_2O nanocrystals as a novel integrated photocatalytic adsorbent. Journal of Materials Chemistry A. 2014;**2**:14563-14570. DOI: 10.1039/c4ta02459a

[29] Mishra M, Chun DM. α-Fe_2O_3 as a photocatalytic material: A review. Applied Catalysis A: General. 2015;**498**:126-141. DOI: 10.1016/j. apcata.2015.03.023

[30] Sajjadia SH, Goharshadi EK. Highly monodispersed hematite cubes for removal of ionic dyes. Journal of Environmental Chemical Engineering. 2017;**5**:1096-1106. DOI: 10.1016/j. jece.2017.01.035

[31] Yuan K, Taylor SD, Powell BA, Becker U. An *ab initio* study of the adsorption of Eu^{3+}, Pu^{3+}, Am^{3+}, and Cm^{3+} hydroxide complexes on hematite (001) surface: Role of magnetism on adsorption. Surface Science. 2017;**664**:120-128. DOI: 10.1016/j. susc.2017.06.007

[32] Noerpel MR, Lenhart JJ. The impact of particle size on the adsorption of citrate to hematite. Journal of Colloid and Interface Science. 2015;**460**:36-46. DOI: 10.1016/j.jcis.2015.08.028

[33] Chen YH, Lin CC. Effect of nano-hematite morphology on photocatalytic activity. Physics and Chemistry of Minerals. 2014;**41**:727-736. DOI: 10.1007/s00269-014-0686-9

[34] Asoufi HM, Al-Antary TM, Awwad AM. Green route for synthesis hematite (α-Fe_2O_3) nanoparticles: Toxicity effect on the green peach aphid, *Myzus persicae* (Sulzer). Environmental Nanotechnology, Monitoring & Management. 2018;**9**:107-111. DOI: 10.1016/j.enmm.2018.01.004

[35] Abu-Dief AM, Abdel-Fatah SM. Development and functionalization of magnetic nanoparticles as powerful and green catalysts for organic synthesis. Beni-Suef University Journal of Basic and Applied Sciences. 2018;**7**:55-67. DOI: 10.1016/j.bjbas.2017.05.008

[36] Kefeni KK, Msagati TAM, Nkambule TTI, Mamba BB. Synthesis and application of hematite nanoparticles for acid mine drainage treatment. Journal of Environmental Chemical Engineering. 2018;**6**: 1865-1874. DOI: 10.1016/j. jece.2018.02.037

[37] Mansour H, Bargougui R, Autret-Lambert C, Gadri A, Ammar S. Co-precipitation synthesis and characterization of tin-doped α-Fe_2O_3 nanoparticles with enhanced photocatalytic activities. Journal of Physics and Chemistry of Solids. 2018;**114**:1-7. DOI: 10.1016/j. jpcs.2017.11.013

[38] Fereshteh Z, Salavati-Niasari M. Effect of ligand on particle size and morphology of nanostructures synthesized by thermal decomposition of coordination compounds. Advances in Colloid and Interface Science. 2017;**243**:86-104. DOI: 10.1016/j.cis.2017.03.001

[39] Darezereshki E. One-step synthesis of hematite (α-Fe_2O_3) nano-particles by direct thermal-decomposition of maghemite. Materials Letters. 2011;**65**:642-645. DOI: 10.1016/j. matlet.2010.11.030

[40] Al-Gaashani R, Radiman S, Tabet N, Daud AR. Rapid synthesis and

optical properties of hematite (α-Fe_2O_3) nanostructures using a simple thermal decomposition method. Journal of Alloys and Compounds. 2013;**550**: 395-401. DOI: 10.1016/j.jallcom. 2012.10.150

[41] Wang J, Shao X, Zhang Q, Tiana G, Ji X, Bao W. Preparation of mesoporous magnetic Fe_2O_3 nanoparticle and its application for organic dyes removal. Journal of Molecular Liquids. 2017;**248**:13-18. DOI: 10.1016/j. molliq.2017.10.026

[42] Yang X, Xia L, Li J, Dai M, Yang G, Song S. Adsorption of As(III) on porous hematite synthesized from goethite concentrate. Chemosphere. 2017;**169**:188-193. DOI: 10.1016/j. chemosphere.2016.11.061

[43] Nithya VD, Arul NS. Review on α-Fe_2O_3 based negative electrode for high performance supercapacitors. Journal of Power Sources. 2016; **327**:297-318. DOI: 10.1016/j.jpowsour. 2016.07.033

[44] Boumaza S, Kabir H, Gharbi I, Belhadi A, Trari M. Preparation and photocatalytic H_2-production on α-Fe_2O_3 prepared by sol-gel. International Journal of Hydrogen Energy. 2018;**43**:3424-3430. DOI: 10.1016/j. ijhydene.2017.07.227

[45] Xiao Q, Sun Y, Zhang J, Li Q. Size-dependent of chromium (VI) adsorption on nano α-Fe_2O_3 surface. Applied Surface Science. 2015;**356**: 18-23. DOI: 10.1016/j.apsusc.2015. 08.005

[46] Mathevula LE, Noto LL, Mothudi BM, Chithambo M, Dhlamini MS. Structural and optical properties of sol-gel derived α-Fe_2O_3 nanoparticles. Journal of Luminescence. 2017;**192**: 879-887. DOI: 10.1016/j.jlumin.2017.07.055

[47] Raja K, Jaculine MM, Jose M, Verma S, Prince AAM, Ilangovan K, et al. Sol-gel synthesis and characterization of α-Fe_2O_3 nanoparticles. Superlattices and Microstructures. 2015;**86**:306-312. DOI: 10.1016/j.spmi.2015.07.044

[48] Lian S, Li H, He X, Kang Z, Liu Y, Lee ST. Hematite homogeneous core/ shell hierarchical spheres: Surfactant-free solvothermal preparation and their improved catalytic property of selective oxidation. Journal of Solid State Chemistry. 2012;**185**:117-123. DOI: 10.1016/j.jssc.2011.11.003

[49] Sun T, Zhu Y, Qi C, Ding G, Chen F, Wu J. α-Fe_2O_3 nanosheet-assembled hierarchical hollow mesoporous microspheres: Microwave-assisted solvothermal synthesis and application in photocatalysis. Journal of Colloid and Interface Science. 2016;**463**:107-117. DOI: 10.1016/j.jcis.2015.10.038

[50] Bouhjar F, Mollar M, Chourou ML, Marí B, Bessaïs B. Hydrothermal synthesis of nanostructured Cr-doped hematite with enhanced photoelectrochemical activity. Electrochimica Acta. 2018;**260**:838-846. DOI: 10.1016/j. electacta.2017.12.049

[51] Colombo C, Palumbo G, Iorio E, Song X, Jiang Z, Liu Q, et al. Influence of hydrothermal synthesis conditions on size, morphology and colloidal properties of hematite nanoparticles. Nano-Structures & Nano-Objects. 2015;**2**:19-27. DOI: 10.1016/j. nanoso.2015.07.004

[52] Wang F, Qin XF, Meng YF, Guo ZL, Yang LX, Ming YF. Hydrothermal synthesis and characterization of α-Fe_2O_3 nanoparticles. Materials Science in Semiconductor Processing. 2013;**16**:802-806. DOI: 10.1016/j. mssp.2012.12.029

[53] Sengupta A, Mallick S, Bahadur D. Tetragonal nanostructured zirconia modified hematite mesoporous composite for efficient adsorption

of toxic cations from wastewater. Journal of Environmental Chemical Engineering. 2017;5:5285-5292. DOI: 10.1016/j.jece.2017.10.002

[54] Dong Y, Xing L, Hu F, Umar A, Wu X. Efficient removal of organic dyes molecules by grain-like α-Fe_2O_3 nanostructures under visible light irradiation. Vacuum. 2018;**150**:35-40. DOI: 10.1016/j.vacuum.2018.01.023

[55] Chen Y, Li F. Kinetic study on removal of copper(II) using goethite and hematite nano-photocatalysts. Journal of Colloid and Interface Science. 2010;**347**:277-281. DOI: 10.1016/j.jcis.2010.03.050

[56] Cheng X, Jiang J, Jin C, Lin C, Zeng Y, Zhang Q. Cauliflower-like α-Fe_2O_3 microstructures: Toluene-water interfaceassisted synthesis, characterization, and applications in wastewater treatment and visible-light photocatalysis. Chemical Engineering Journal. 2014;**236**:139-148. DOI: 10.1016/j.cej.2013.09.089

[57] Liu Y, Yu C, Dai W, Gao X, Qian H, Hua Y, et al. One-pot solvothermal synthesis of multi-shelled α-Fe_2O_3 hollow spheres with enhanced visible-light photocatalytic activity. Journal of Alloys and Compounds. 2013;**551**: 440-443. DOI: 10.1016/j.jallcom.2012.11.047

[58] Rakhshaee R, Darvazeh J. Comparing performance of three forms of hematite in fixed bed reactor for a photocatalytic decolorization: Experimental design, model fitting and optimization of conditions. Process Safety and Environmental Protection. 2017;**107**:122-137. DOI: 10.1016/j.psep.2017.02.006

[59] Mahmoodi NM, Abdi J, Oveisi M, Asli MA, Vossoughi M. Metal-organic framework (MIL-100 (Fe)): Synthesis, detailed photocatalytic dye degradation ability in colored textile wastewater and recycling. Materials Research Bulletin. 2018;**100**:357-366. DOI: 10.1016/j.materresbull.2017.12.033

[60] Shen L, Liang R, Wu L. Strategies for engineering metal-organic frameworks as efficient photocatalysts. Chinese Journal of Catalysis. 2015;**36**:2071-2088. DOI: 10.1016/S1872-2067(15)60984-6

[61] Zhang H, Nai J, Yu L, Lou XWD. Metal-organic-framework-based materials as platforms for renewable energy and environmental applications. Joule. 2017;**1**:77-107. DOI: 10.1016/j.joule.2017.08.008

[62] Valizadeh B, Nguyen TN, Stylianou KC. Shape engineering of metal-organic frameworks. Polyhedron. 2018;**145**:1-15. DOI: 10.1016/j.poly.2018.01.004

[63] Ban J, Xu G, Zhang L, Lin H, Sun Z, Lv Y, et al. Mesoporous ZnO microcube derived from a metal-organic framework as photocatalyst for the degradation of organic dyes. Journal of Solid State Chemistry. 2017;**256**:151-157. DOI: 10.1016/j.jssc.2017.09.002

[64] Gao Q, Xu J, Bu X. Recent advances about metal-organic frameworks in the removal of pollutants from wastewater. Coordination Chemistry Reviews. In Press. DOI: 10.1016/j.ccr.2018.03.015

[65] He Y, Chen F, Li B, Qian G, Zhou W, Chen B. Porous metal-organic frameworks for fuel storage. Coordination Chemistry Reviews. 2018;**373**:167-198. DOI: 10.1016/j.ccr.2017.10.002

[66] Zhang J, Chen Z. Metal-organic frameworks as stationary phase for application in chromatographic separation. Journal of Chromatography A. 2017;**1530**:1-18. DOI: 10.1016/j.chroma.2017.10.065

[67] Qiu J, Zhang X, Feng Y, Zhang X, Wang H, Yao J. Modified metal-organic frameworks as photocatalysts. Applied Catalysis B: Environmental. 2018;**231**:317-342. DOI: 10.1016/j.apcatb.2018.03.039

[68] Ramezanalizadeh H, Manteghi F. Synthesis of a novel MOF/$CuWO_4$ heterostructure for efficient photocatalytic degradation and removal of water pollutants. Journal of Cleaner Production. 2018;**172**:2655-2666. DOI: 10.1016/j.jclepro.2017.11.145

[69] Wen J, Fang Y, Zeng G. Progress and prospect of adsorptive removal of heavy metal ions from aqueous solution using metal-organic frameworks: A review of studies from the last decade. Chemosphere. 2018;**201**:627-643. DOI: 10.1016/j.chemosphere.2018.03.047

[70] Hashemi B, Zohrabi P, Raza N, Kim K. Metal-organic frameworks as advanced sorbents for the extraction and determination of pollutants from environmental, biological, and food media. Trends in Analytical Chemistry. 2017;**97**:65-82. DOI: 10.1016/j.trac.2017.08.015

[71] Khan NA, Jhung SH. Adsorptive removal and separation of chemicals with metal-organic frameworks: Contribution of π-complexation. Journal of Hazardous Materials. 2017;**325**:198-213. DOI: 10.1016/j.jhazmat.2016.11.070

[72] Ahmed I, Jhung SH. Applications of metal-organic frameworks in adsorption/separation processes via hydrogen bonding interactions. Chemical Engineering Journal. 2017;**310**:197-215. DOI: 10.1016/j.cej.2016.10.115

[73] Fan Y, Zhang S, Qin S, Li X, Qi S. An enhanced adsorption of organic dyes onto NH_2 functionalization titanium-based metal-organic frameworks and the mechanism investigation. Microporous and Mesoporous Materials. 2018;**263**:120-127. DOI: 10.1016/j.micromeso.2017.12.016

[74] Sarker M, Song JY, Jhung SH. Adsorption of organic arsenic acids from water over functionalized metal-organic frameworks. Journal of Hazardous Materials. 2017;**335**:162-169. DOI: 10.1016/j. jhazmat.2017.04.044

[75] Zhang Q, Cui Y, Qian G. Goal-directed design of metal-organic frameworks for liquid phase adsorption and separation. Coordination Chemistry Reviews. In Press. DOI: 10.1016/j. ccr.2017.10.028

[76] Pi Y, Li X, Xia Q , Wu J, Li Y, Xiao J, et al. Adsorptive and photocatalytic removal of persistent organic pollutants (POPs) in water by metal-organic frameworks (MOFs). Chemical Engineering Journal. 2018;**337**:351-371. DOI: 10.1016/j.cej.2017.12.092

[77] Sharma VK, Feng M. Water depollution using metal-organic frameworks-catalyzed advanced oxidation processes: A review. Journal of Hazardous Materials. In Press. DOI: 10.1016/j.jhazmat.2017.09.043

[78] Wang S, Wang X. Multifunctional metal-organic frameworks for photocatalysis. Small. 2015;**11**:3097-3112. DOI: 10.1002/smll.201500084

[79] Seetharaj R, Vandana PV, Arya P, Ma S. Dependence of solvents, pH, molar ratio and temperature in tuning metal organic framework architecture. Arabian Journal of Chemistry. In Press. DOI: 10.1016/j.arabjc.2016.01.003

[80] Beg S, Rahman M, Jain A, Saini S, Midoux P, Pichon C, et al. Nanoporous metal organic frameworks as hybrid polymer-metal composites for drug delivery and biomedical applications.

Drug Discovery Today. 2017;**22**:625-637. DOI: 10.1016/j.drudis.2016.10.001

[81] Wang Y, Li L, Yan L, Cao L, Dai P, Gu X, et al. Continuous synthesis for zirconium metal-organic frameworks with high quality and productivity via microdroplet flow reaction. Chinese Chemical Letters. 2018;**29**:849-853. DOI: 10.1016/j. cclet.2017.09.057

[82] Panahi L, Naimi-Jamal MR, Mokhtari J, Morsali A. Mechanochemically synthesized nanoporous metal-organic framework $Cu_2(BDC)_2(DABCO)$: An efficient heterogeneous catalyst for preparation of carbamates. Microporous and Mesoporous Materials. 2017;**244**:208-217. DOI: 10.1016/j. micromeso.2016.10.031

[83] Majedi A, Davar F, Abbasi AR. Metal-organic framework materials as nano photocatalyst. International Journal of Nano Dimension. 2016;**7**:1-14. DOI: 10.7508/ijnd.2016.01.001

[84] Gangu KK, Maddila S, Mukkamala SB, Jonnalagadda SB. A review on contemporary metal-organic framework materials. Inorganica Chimica Acta. 2016;**446**:61-74. DOI: 10.1016/j. ica.2016.02.062

[85] Ren J, Dyosiba X, Musyoka NM, Langmi HW, Mathe M, Liao S. Review on the current practices and efforts towards pilot-scale production of metal-organic frameworks (MOFs). Coordination Chemistry Reviews. 2017;**352**:187-219. DOI: 10.1016/j. ccr.2017.09.005

[86] Pirzadeh K, Ghoreyshi AA, Rahimnejad M, Mohammadi M. Electrochemical synthesis, characterization and application of a microstructure $Cu_3(BTC)_2$ metal organic framework for CO_2 and CH_4 separation. Korean Journal of Chemical Engineering. 2018;**35**:974-983. DOI: 10.1007/s11814-017-0340-6

[87] Van Assche TRC, Denayer JFM. Fabrication and separation performance evaluation of a metal-organic framework based microseparator device. Chemical Engineering Science. 2013;**95**:65-72. DOI: 10.1016/j. ces.2013.03.006

[88] Yang H, Liu X, Song X, Yang T, Liang Z, Fan C. In situ electrochemical synthesis of MOF-5 and its application in improving photocatalytic activity of BiOBr. Transactions of Nonferrous Metals Society of China. 2015;**25**:3987-3994. DOI: 10.1016/ S1003-6326(15)64047-X

[89] Giménez-Marqués M, Hidalgo T, Serre C, Horcajada P. Nanostructured metal-organic frameworks and their bio-related applications. Coordination Chemistry Reviews. 2016;**307**:342-360. DOI: 10.1016/j.ccr.2015.08.008

[90] Wang C, Du X, Li J, Guo X, Wang P, Zhang J. Photocatalytic Cr(VI) reduction in metal-organic frameworks: A mini-review. Applied Catalysis B: Environmental. 2016;**193**:198-216. DOI: 10.1016/j.apcatb.2016.04.030

[91] Klimakow M, Klobes P, Rademann K, Emmerling F. Characterization of mechanochemically synthesized MOFs. Microporous and Mesoporous Materials. 2012;**154**:113-118. DOI: 10.1016/j. micromeso.2011.11.039

[92] Masoomi MY, Stylianou KC, Morsali A, Retailleau P, Maspoch D. Selective CO_2 capture in metal-organic frameworks with azine-functionalized pores generated by mechanosynthesis. Crystal Growth & Design. 2014;**14**:2092-2096. DOI: 10.1021/ cg500033b

[93] Chen Y, Xiao J, Lv D, Huang T, Xu F, Sun X, et al. Highly efficient mechanochemical synthesis of an indium based metalorganic framework with excellent water stability.

Chemical Engineering Science. 2017;**158**:539-544. DOI: 10.1016/j.ces.2016.11.009

[94] Seoane B, Castellanos S, Dikhtiarenko A, Kapteijn F, Gascon J. Multi-scale crystal engineering of metal organic frameworks. Coordination Chemistry Reviews. 2016;**307**:147-187. DOI: 10.1016/j.ccr.2015.06.008

[95] Zhu J, Li P, Guo W, Zhao Y, Zou R. Titanium-based metal-organic frameworks for photocatalytic applications. Coordination Chemistry Reviews. 2015;**359**:80-101. DOI: 10.1016/j.ccr.2017.12.013

[96] Abazari R, Mahjoub AR. Ultrasound-assisted synthesis of zinc(II)-based metal organic framework nanoparticles in the presence of modulator for adsorption enhancement of 2,4-dichlorophenol and amoxicillin. Ultrasonics Sonochemistry. 2018;**42**:577-584. DOI: 10.1016/j.ultsonch.2017.12.027

[97] Masoomi MY, Bagheri M, Morsali A. Porosity and dye adsorption enhancement by ultrasonic synthesized Cd(II) based metal-organic framework. Ultrasonics Sonochemistry. 2017;**37**:244-250. DOI: 10.1016/j.ultsonch.2017.01.018

[98] Abdollahi N, Masoomi MY, Morsali A, Junk PC, Wang J. Sonochemical synthesis and structural characterization of a new Zn(II) nanoplate metal-organic framework with removal efficiency of Sudan red and Congo red. Ultrasonics Sonochemistry. 2018;**45**:50-56. DOI: 10.1016/j.ultsonch.2018.03.001

[99] Batten MP, Rubio-Martinez M, Hadley T, Carey K, Lim K, Polyzos A, et al. Continuous flow production of metal-organic frameworks. Current Opinion in Chemical Engineering. 2015;**8**:55-59. DOI: 10.1016/j.coche.2015.02.001

[100] Stock N, Biswas S. Synthesis of metal-organic frameworks (mofs): Routes to various mof topologies, morphologies, and composites. Chemical Reviews. 2012;**112**:933-969. DOI: 10.1021/cr200304e

[101] He J, Li J, Du W, Han Q, Wang Z, Li M. A mesoporous metal-organic framework: Potential advances in selective dye adsorption. Journal of Alloys and Compounds. 2018;**750**:360-367. DOI: 10.1016/j.jallcom.2018.03.393

[102] Zhang J, Li F, Sun Q. Rapid and selective adsorption of cationic dyes by a unique metal-organic framework with decorated pore surface. Applied Surface Science. 2018;**440**:1219-1226. DOI: 10.1016/j.apsusc.2018.01.258

[103] Sanram S, Boonmak J, Youngme S. Ni(II)-metal-organic frameworks based on 1,4-phenylenedipropionic acid: Solvothermal syntheses, structures, and photocatalytic properties. Polyhedron. 2016;**119**:151-159. DOI: 10.1016/j.poly.2016.08.044

[104] Gao Y, Li S, Li Y, Yao L, Zhang H. Accelerated photocatalytic degradation of organic pollutant over metal-organic framework MIL-53(Fe) under visible LED light mediated by persulfate. Applied Catalysis B: Environmental. 2017;**202**:165-174. DOI: 10.1016/j.apcatb.2016.09.005

[105] Gao Y, Yu G, Liu K, Deng S, Wang B, Huang J, et al. Integrated adsorption and visible-light photodegradation of aqueous clofibric acid and carbamazepine by a Fe-based metal-organic framework. Chemical Engineering Journal. 2017;**330**:157-165. DOI: 10.1016/j.cej.2017.06.139

[106] Araya T, Chen C, Jia M, Johnson D, Li R, Huang Y. Selective degradation of organic dyes by a resin modified Fe-based metal-organic framework under visible light irradiation. Optical

Materials. 2017;**64**:512-523. DOI: 10.1016/j.optmat.2016.11.047

[107] Wang D, Jia F, Wang H, Chen F, Fang Y, Dong W, et al. Simultaneously efficient adsorption and photocatalytic degradation of tetracycline by Fe-based MOFs. Journal of Colloid and Interface Science. 2018;**519**:273-284. DOI: 10.1016/j.jcis.2018.02.067

[108] Abdpour S, Kowsari E, Moghaddam MRA. Synthesis of MIL-100(Fe)@MIL-53(Fe) as a novel hybrid photocatalyst and evaluation photocatalytic and photoelectrochemical performance under visible light irradiation. Journal of Solid State Chemistry. 2018;**262**:172-180. DOI: 10.1016/j.jssc.2018.03.018

[109] Lan Y, Lu Y, Ren Z. Mini review on photocatalysis of titanium dioxide nanoparticles and their solar applications. Nano Energy. 2013;**2**:1031-1045. DOI: 10.1016/j.nanoen.2013.04.002

[110] Patil RA, Devan RS, Liou Y, Ma Y. Efficient electrochromic smart windows of one-dimensional pure brookite TiO_2 nanoneedles. Solar Energy Materials & Solar Cells. 2016;**147**:240-245. DOI: 10.1016/j.solmat.2015.12.024

[111] Verbruggen SW. TiO_2 photocatalysis for the degradation of pollutants in gas phase: From morphological design to plasmonic enhancement. Journal of Photochemistry and Photobiology C: Photochemistry Reviews. 2015;**24**:64-82. DOI: 10.1016/j.jphotochemrev.2015.07.001

[112] Etacheri V, Valentin C, Schneider J, Bahnemann D, Pillai SC. Visible-light activation of TiO_2 photocatalysts: Advances in theory and experiments. Journal of Photochemistry and Photobiology C: Photochemistry Reviews. 2015;**25**:1-29. DOI: 10.1016/j.jphotochemrev.2015.08.003

[113] Singh S, Mahalingam H, Singh PK. Polymer-supported titanium dioxide photocatalysts for environmental remediation: A review. Applied Catalysis A: General. 2013;**462-463**:178-195. DOI: 10.1016/j.apcata.2013.04.039

[114] Singh R, Dutta S. A review on H_2 production through photocatalytic reactions using TiO_2/TiO_2-assisted catalysts. Fuel. 2018;**220**:607-620. DOI: 10.1016/j.fuel.2018.02.068

[115] Wang Y, He Y, Lai Q, Fan M. Review of the progress in preparing nano TiO_2: An important environmental engineering material. Journal of Environmental Sciences. 2014;**26**: 2139-2177. DOI: 10.1016/j.jes.2014.09.023

[116] Khaki MRD, Shafeeyan MS, Raman AAA, Daud WMAW. Application of doped photocatalysts for organic pollutant degradation—A review. Journal of Environmental Management. 2017;**198**:78-94. DOI: 10.1016/j.jenvman.2017.04.099

[117] Cheng G, Xu F, Xiong J, Wei Y, Stadler FJ, Chen R. A novel protocol to design TiO_2-Fe_2O_3 hybrids with effective charge separation efficiency for improved photocatalysis. Advanced Powder Technology. 2017;**28**:665-670. DOI: 10.1016/j.apt.2016.12.004

[118] Cheng L, Qiu S, Chen J, Shao J, Cao S. A practical pathway for the preparation of Fe_2O_3 decorated TiO_2 photocatalyst with enhanced visible-light photoactivity. Materials Chemistry and Physics. 2017;**190**:53-61. DOI: 10.1016/j.matchemphys.2017.01.001

[119] Chang N, Zhang H, Shi M, Li J, Shao W, Wang H. Metal-organic framework templated synthesis of TiO_2@MIL-101 coreshell architectures for high-efficiency adsorption and photocatalysis. Materials Letters.

2017;**200**:55-58. DOI: 10.1016/j.matlet.2017.04.099

[120] Huang J, Song H, Chen C, Yang Y, Xu N, Ji X, et al. Journal of Environmental Chemical Engineering. 2017;**5**:2579-2585. DOI: 10.1016/j.jece.2017.05.012

[121] Dal'Toé ATO, Colpani GL, Padoin N, Fiori MA, Soares C. Lanthanum doped titania decorated with silver plasmonic nanoparticles with enhanced photocatalytic activity under UV-visible light. Applied Surface Science. 2018;**441**:1057-1071. DOI: 10.1016/j.apsusc.2018.01.291

[122] Colpani GL, Zanetti JT, Cecchin F, Dal'Toé A, Fiori MA, Moreira RFPM, et al. Carboxymethyl-β-cyclodextrin functionalization of TiO_2 doped with lanthanum: Characterization and enhancement of photocatalytic activity. Catalysis Science & Technology. 2018;**8**:2636-2647. DOI: 10.1039/C7CY02115A

Chapter 5

Effect of Biochar Amendments on the Sorption and Desorption Herbicides in Agricultural Soil

Kassio Ferreira Mendes, Ananias Francisco Dias Júnior, Vanessa Takeshita, Ana Paula Justiniano Régo and Valdemar Luiz Tornisielo

Abstract

Improved understanding of herbicide destinations, effects, and environmental risks through worldwide studies is crucial to minimizing impacts to nontarget organisms, especially in tropical regions rich in biodiversity. In recent years, there has been widespread international concern about the toxic effects of herbicides on humans, faunas, and native floras. Therefore, the adoption of agricultural practices that minimize the environmental effects of herbicides has been frequently studied, for example, the addition of biochar in agricultural soils. Biochar can be defined as the by-product of a thermal process conducted under low oxygen or oxygen-free conditions (pyrolysis) to convert plant biomass to biofuels, where biochar is the solid product of pyrolysis. The addition of biochar to the soil can easily potentiate the herbicide retention process, which, in addition to contributing positively to the reduction of chemical contaminants in the environment, may exert negative effects on herbicide behavior and the efficacy of these products on weed control. Thus, this chapter will present the general characteristics of biochar, as well as the impact of this material on sorption-desorption of herbicides in the soil.

Keywords: bioavailability, black carbon, contaminated soils, pyrolysis, retention process

1. Introduction

Improved understanding of herbicide fate, effects, and environmental risks through worldwide studies is crucial to minimize impacts to nontarget organisms, especially in tropical regions rich in biodiversity [1]. In recent years, according to Yavari et al. [2], there has been widespread international concern about the toxic effects of herbicides on humans, faunas, and native floras. Therefore, the adoption of agricultural practices that minimize the environmental effects of herbicides has been frequently studied, for example, the addition of biochar (charcoal or black carbon) to agricultural soils.

Biochar can be defined as the by-product of a thermal process conducted under low oxygen or oxygen-free conditions (pyrolysis) to convert plant biomass to biofuels [3, 4], where the biochar is the solid product of pyrolysis [5]. Pyrolysis or carbonization is a process that involves the application of heat to the biomass in order to concentrate the heat and collect the by-products. It is an interesting alternative, mainly for the treatment of residues, in general of the biomass, aiming its direct application in the soil or composing the compost. The main difference between pyrolysis and incineration is that by pyrolysis it is possible to recover condensable gases (pyroligneous or bio-oil) and those that do not condense, but with high combustion power providing additional energy for the processing unit. Added to this, its contribution is notable for the fact that it minimizes the emissions of greenhouse gases (GHGs) inherent in current agriculture.

Initially, biochar, described as "Terra-Preta de Índio" by Sombroek [6], found in Amazonian soils, and formed by the anthropogenic addition of ceramics and artifacts, has pyrogenic carbon molecules that undergo partial carbonization and are more stable than other forms [7, 8]. Studies of the material provided the basis for agricultural use because of its properties and its benefits to the soil. Since then, biochar has been used to mitigate agronomic problems [9]. Biochar has also been studied as an alternative in the remediation of chemical contaminants in the soil; its use has implicated in the behavior and efficacy of herbicides in the control of weeds and in the environmental impact of these products.

In addition to being found in nature, biochar can also be produced artificially, as previously described and the factors related to the pyrolytic process, such as temperature, heating rate, and pressure, can alter the recovery amounts of each final product, the values of energy of the bio-oils and the physicochemical properties of biochar [10], as well as the types of materials used in the firing can present different answers regarding these characteristics.

Biochar has been used in agricultural fields with positive effects on the soil microbiota [11], changes in soil properties, increasing surface area, pH, C/N ratio [12], nutrient cycling in the soil, increase of available water in the soil for the plants, soil organic matter (SOM) construction, reduction of soil bulk density, C sequestration, and reduction of herbicide transport to surface and subsurface waters [13]. This carbonaceous material can also stabilize heavy metals and decrease their release at levels toxic to the soil [2], being an important alternative in soils with water deficit and nutrient deficiency. However, biochar production by pyrolysis with an incomplete combustion process can be considered as a producer of pollution, containing inorganic molecules, heavy metals, and others that can be harmful to the environment [14].

The use of biochar in agricultural soils as fertilizer and soil conditioner has been more exploited; however, little is known of the effect of this material on soil contaminated with herbicides and also its effect on weed control. The fate and environmental behavior of herbicides as well as their effectiveness when applied directly to the soil, such as preemergent herbicides with residual action, are strongly influenced by the retention binding with soil colloidal particles and the organic carbon (OC) content [15]. Therefore, the addition of biochar to the soil can easily potentiate the herbicide retention process, which, in addition to contributing positively to the reduction of chemical contaminants in the environment, may exert negative effects on herbicide behavior and the effectiveness of these products in weed control. Thus, this chapter will present the general characteristics of biochar, as well as the impact of this material on sorption-desorption of herbicides in the soil.

2. General characteristics of biochar

In an approach on the issue, Brito [16] presents several factors that can act on the nature and yields of pyrolysis products. These factors are as follows:

a. Factors related to the nature of the raw material: elemental composition, density, grain size, mineral content, composition of the three main polymers (cellulose, hemicelluloses, and lignin), calorific value, and mechanical strength.

b. Process factors: final temperature, pressure, residence time in the heating zone, heating rate, thermal fluxes and the heat transfer coefficients resulting from the heating rate, chemical or thermal pretreatments in the biomass.

However, here are described the aspects related to the use and efficiency of biochar in the soil, aiming to increase the properties of the set.

2.1 Surface area and porosity

In general, the action of heat on biomass via pyrolysis (>500°C) results in a more graphitized (aromatic) material—biochar, with greater specific surface area and reduced abundance of surface functional groups. In contrast, lower pyrolysis temperatures (<400°C) produce biochars with low specific surface area, which implies partitioning and specific interactions with functional groups on the surface of the biochar [17]. Porosity is another property affected. Higher temperatures result in a more porous biochar when compared to those obtained under lower pyrolysis temperatures [17, 18]. According to the authors, the increase in porosity is accompanied by an increase in the proportion of micropores that contribute significantly to the sorption power of these carbonaceous materials. For example, **Figure 1** shows an image of scanning electron microscope (SEM) of the biochar derived of *Eucalyptus* spp. The biochar exhibits a rough and irregular surface that is characteristic of the material.

In addition, it is well known that desorption is affected by the extent of herbicide absorption in the biochar, controlling its bioavailability in agricultural soils. In general, higher pyrolysis temperatures lead to higher hysteretic desorption processes [17]. One of the intrinsic properties of biochar is its sorption capacity. It may influence other properties, such as mechanical resistance. In India, Japan, and some European countries, its high sorption capacity has been used to capture radioactive elements in the soil, such as Radon (Rn), for example [18].

According to Andrade and Della Lucia [19], this sorption is related to the high porosity of biochar, given by the difference between its actual specific mass and its apparent specific mass. The authors also stated that the porosity of the biochar is linked to the final temperature of the applied pyrolysis, but also add to the existence of the effects of the density of the raw material that gave rise to it.

Figure 1.
Scanning electron microscope (SEM) image of biochar derived of Eucalyptus spp. Magnification is 1000×.

According to Maia et al. [8], the chemical composition of the biomass is highly variable according to the botanical species and the biomass part (leaves, branches, wood, bagasse, residues from the extraction of vegetable oil, among others) and thus significantly influences the products obtained after pyrolysis. Added to this, they provide anatomically distinct biochar (porosity, grain size, among others).

Yu et al. [20] observed higher sorption of diuron in the biochar produced at 850°C than in another produced at 450°C. According to Sharma et al. [21], the elevation of the pyrolysis temperature in the biochar production can raise the retention potential of organic contaminants when applied to the soil; however, the pyrolysis at 400°C decreased the surface area of the biochar particle. Chen et al. [22] indicated that at the temperature of 700°C, the biochar presents half of the specific surface area. The temperature of the process determines the type of carbon present in the biochar, with the reaction time being less decisive [23].

Thus, porosity constitutes one of the most relevant characteristics for biochar aiming at its application in agricultural soils. Yu et al. [20] studied the sorption capacity of herbicidal agents in soils with and without the incorporation of biochar produced from different species and pyrolysis temperatures. The results obtained by the authors showed better results for the soils that received the biochar obtained at the highest temperature, regardless of the biomass used.

2.2 Apparent specific mass and true density

The density of feedstock is of great importance in obtaining the biochar, since for the same mass, one desires good yields in conversion of the biochar, provided that the density of the feedstock is high. In practical terms, the higher the density of the source material, the higher the density of the biochar after the pyrolysis. Some studies report that the increase in density is associated with the increase of the lignin content, which implies higher yield in biochar. According to Pétroff and Doat [24], lignin is rich in carbon and thus favors conversion.

The true density is another variable that makes up the biochar, that is, the apparent specific mass discounting the volume of the internal porosity. When the true density is related to the apparent density, the porosity is measured. The porosity as mentioned above (Section 2.1) is the measurement of empty space, constituting an intrinsic characteristic of biochar with direct influence on its hygroscopicity, reactivity, combustion performance, and sorption capacity. The true density is dependent on the temperature of the pyrolysis used, where the higher the temperature, the higher the true density and, consequently, the greater the porosity of the biochar produced. In order to determine the porosity of the porous surface, the porous surface of the porous surface of the porous surface of the porous surface of the porous surface of the porous surface of the porous surface of the porous porous surface of the porous porous porous surface of the porous porous porous pores. Thus, calculated as a function of apparent density and true density, the porosity will increase as a function of the apparent specific mass of the biochar [25].

2.3 pH

The addition of biochar to the soil, in granulometry similar to the fractions that make up the sand, silt, and clay, can alter the limits of consistency, improve water retention capacity, increase pH, and contribute to soil structure improvement [17]. This fact is attributed to the great presence of positive charges at the ends of the carbon chains of the biochar. The subsidized hypothesis for this is that the higher the pyrolysis temperature, the higher the fixed carbon content, and consequently the greater the presence of positive charges in the structure obtained from the biochar.

Clay and Malo [26] found that the biochars of corn and *Panicum virgatum*, produced at high temperatures (>650°C), regardless of processing time, were very alkaline (pH > 9). In processes with lower temperatures (<550°C), the materials had pH < 5.

However, Yang et al. [27] considered the pyrolysis temperature more relevant in the production of biochar for the treatment efficiency of contaminated soils than the material used. In studies comparing the pyrolysis temperature of the wood biochar of *Pinus radiata*, the herbicide terbuthylazine was more sorbed in the biochar obtained at 700° C than those obtained at 350°C [28]. In contrast, Li et al. [29], studying the leaching of 2,4-D and acetochlor, found a reduction in leaching and amplification in herbicide efficacy with biochars produced at low-temperature (350°C) pyrolysis.

In practice, the biochars obtained by higher pyrolysis temperatures have a higher capacity to raise the pH of the soils that were initially incorporated.

2.4 Relationship C:N and H:C and minerals

The biochar obtained from the pyrolysis biomass generally presents low content of nitrogen and hydrogen, which results in a high C:N and H:C ratio [30]. Oxygen is the second most abundant element of the material and its content is inversely related to the final pyrolysis temperature applied. There are also ashes, which come from the mineral elements mainly from bark. In the ashes, potassium, calcium, phosphorus, and sodium predominate. The composition of the ashes is strongly related to the chemistry of the soils where the original biomass was developed. These properties give the biochar a great capacity of persistence in the soil. In studies of dating of biochar fragments (coal) found in soils, it is common to observe samples with thousands of years [31]. Some studies also suggest the positive effect of the coals on physical water properties of soils, increasing their retention capacity and humidity [32]. In general, the pyrolysis gives rise to the C:N and H:C ratio in relation to the raw material that gave rise to the biochar.

2.5 Chemical composition immediately

The immediate chemical composition is formed by the contents of volatile materials, ashes, and fixed carbon. The denominated fraction of volatile material is emitted during the heating of the biochar constituted of molecules of CO, CO_2, and hydrocarbons. Another amount of carbon remains relatively intact, and as it is not eliminated along with the volatile material, it is called the fixed carbon. In practice, the content of volatile material and fixed carbon is determined by heating the biochar at a temperature of around 900°C. Ash is the residue of mineral oxides obtained by the complete combustion of the biochar. The oxidized residue obtained is calculated as the biochar ash content.

Biomass, when subjected to the action of heat, at high temperatures undergoes a process of transformation, in which all its components are extensively modified [33]. Better biochar immediate properties of biochar—higher fixed carbon content and lower volatile and ash content—are associated with high lignin feedstocks for certain pyrolysis conditions. Each temperature range generates a different product, and the final temperature has a great influence on the final characteristics of the biochar.

Singh et al. [34] indicated a difference in the values of the total carbon fixed in biochars of different materials, so that this content was higher in those coming from eucalyptus wood and eucalypt leaves when compared to cattle manure. The higher carbon content influences the sorption capacity of the herbicides. Cabrera et al. [35] verified the effect of different biochars on sorption of bentazone so that the sorption increased according to the organic carbon contents dissolved by the materials. For aminocyclopyrachlor, Cabrera et al. [35] found higher sorption in biochars that contained larger surface area and humification index. Exemplifying the hypothesis that each herbicide interacts with the biochar properties govern its behavior in the environment.

In conclusion, the properties of biochar are attributed to the conditions of pyrolysis, temperature, and time, as well as the raw material used [23]. In Brazil, due to the high availability of wood and its derivatives, besides the agricultural base residues, these materials are the basis of the biomass used as a raw material for conversion into biochar. These materials contain a high proportion of cellulose, hemicelluloses, and lignin, the main compositions being converted into carbon matrix in the formation of the biochar, so that these aggregates and their proportional abundance determine the properties of the biochar produced [2].

3. Impact of biochar on herbicides retention processes in the soil

The environmental performance and herbicide efficacy applied to the soil are strongly influenced by the retention with the soil particles, mainly the OC content [15], which can be potentiated with the addition of biochar. The herbicide-soil interaction is directly related to the physical-chemical properties of the products and the soil, in addition to the environmental conditions. The impacts of biochar on sorption and desorption of herbicides are presented in **Table 1**, according to the surveys of recent years on this subject.

3.1 Sorption herbicides

According to Khorram et al. [14], sorption is the first process that occurs after the addition of the herbicide in the soil. Thus, retention is an important factor that directly affects other processes, such as herbicide transport via leaching, surface runoff, and volatilization, as well as bioavailability and impacts on nontarget organisms [36]. High OC content, higher surface area, and more porous structures result in higher herbicide sorption capacities [23]. In soils amended with biochar, sorption of herbicides can be increased, reducing the risks of contamination and exposure in the ecosystem and human health [37]. Martin et al. [38] found an increase in the Freundlich sorption coefficient (K_f) of atrazine using chicken litter biochar at the dose of 10 t ha^{-1}, when compared to soil without biochar. Tatarková et al. [39] indicated that sorption of MCPA (4-chloro-2-methylphenoxyacetic acid) by biochar without soil and by soil amended with biochar (1.0% m m^{-1}) was 82 and 2.53 times higher than in unamended soil, respectively. For one of the most used herbicides in the world, atrazine, a study demonstrated an increase in the K_f value of 5 for sandy soil and 4.3 times for clayey soil with the 1% amended of wheat biochar [40]. Xu et al. [41] reported increased sorption and K_f values of 1.5 and 3 times in soil, when there was amended soil in 0.1 and 0.5% with rice straw biochar.

The pyrolysis temperature, as previously reported, alters the properties of the biochar, thus altering its relationship with herbicide retention. Using the wheat straw biochar, with a firing temperature of 300°C, Spokas et al. [42] reported the high sorption capacity of the sawdust biochar (5% m m^{-1}) in sandy soil for atrazine and acetochlor, attributed to the high OC content (69%) and specific surface area (1.6 m^2 g^{-1}) of biochar. Hall et al. [43] in glyphosate study obtained higher values of sorption of the herbicide according to the increase in the temperature of burning of the materials, where the highest retention was reported in biochar produced at 900°C.

Feedstock	Herbicide	Retention of herbicide	Effect	Source
Wheat ash (1%)	Diuron	Sorption	Increased sorption (fourfold) in amended soils	Yang and Sheng [48]
Biochar derived from wheat (0.05, 0.5, and 1%)	Diuron	Sorption	Increased sorption (7- to 80-fold) with 1% of biochar	Yang et al. [49]
Biochar derived from *Eucalyptus* spp. (450°C) at 0.1, 0.5, 1.0, 2.0, and 5.0% application rate	Diuron	Sorption	Increased sorption (7- to 80-fold) with 1% of biochar	Yu et al. [20]
Biochar derived from *Eucalyptus* spp. (50°C) at 0.1, 0.2, 0.5, 0.8, and 1.0%	Diuron	Sorption	Increased sorption (5- to 125-fold) in amended soils	Yu et al. [20]
Biochar derived from sawdust	Atrazine and acetochlor	Sorption	Increase of 1.5-fold the K_d for acetochlor. Sorption of atrazine was also increased	Spokas et al. [42]
Sewage of dairy products, 200°C for 4 h, 350°C for 4 h	Atrazine	Sorption	Increase of the sorption in the biochar 200°C in amended soil	Cao et al. [50]
Biochar derived from charcoal (350°C)	Terbuthylazine	Sorption	Increased sorption (2.7-fold) in amended soils	Wang et al. [28]
Biochar derived from sawdust (700°C)	Terbuthylazine	Sorption	Increased sorption (63-fold) in amended soils	Wang et al. [28]
Hardwood sawing (500°C), hardwood (540°C), and wooden pallets (>500°C) at 2%	MCPA and fluometuron	Sorption	Sorption increased by 240–5200%	Cabrera et al. [51]
Biochar produced from chicken bed and wheat straw (400°C)	Fluridone and norflurazon	Sorption	Sorption increased by 24- and 36-fold, respectively	Sun et al. [52]
Paper mill slurry (500°C) at 1–5% and chicken bed (500°C) at 1%	Diuron and atrazine	Sorption	Increase of sorption of diuron in 220–448% and atrazine in 270–515%	Martin et al. [38]

Feedstock	Herbicide	Retention of herbicide	Effect	Source
Eucalyptus sp. (0.1–1%)	Isoproturon	Sorption	Sorption and hysteresis increased	Sopeña et al. [45]
Pallets of wood (>500°C), macadamia nuts (850°C), and hardwood (540°C) at 10%	Aminocyclopyrachlor and bentazone	Sorption	Increased sorption of 18–240% for aminocyclopyrachlor and 13–35% for bentazone	Cabrera et al. [35]
Beechwood (550°C) at 1.5%	Imazamox	Sorption	Increased sorption of <5%	Dechene et al. [53]
Biochar produced from sugarcane bagasse, soybean meal, wood chips, among others	MCPA, nicosulfuron, terbuthylazine, and indaziflam	Sorption	Increased sorption for all herbicides	Trigo et al. [54]
Biochar derived from corn silage (750°C) at 0.5%	Isoproturon	Sorption	The amount of bioavailable herbicide was reduced 10- to 2283-fold in treatment with biochar	Eibisch et al. [55]
Biochar of wood pallets (650°C), wood chips (500°C), and corn bran (490°C)	Aminocyclopyrachlor, picloram, metsulfuron-methyl, oxyfluorfen, and alachlor	Sorption	Aminocyclopyrachlor, metsulfuron-methyl, and picloram showed relatively low sorption; alachlor intermediate sorption and oxyfluorfen heavily sorbed	Hall et al. [56]
Biochar of pecan, cherry, and apple flakes (350, 500, 700, and 900°C) and wooden pallets (350, 500, and 700°C)	Glyphosate	Sorption	Sorption increased according to the pyrolysis temperature (higher at 900°C), depending on concentration	Hall et al. [43]
Biochar derived from sawdust (700°C)	Terbuthylazine	Desorption	Reduced desorption	Wang et al. [28]
Sawing of hardwood (500°C), hardwood (540°C), and wooden pallets (>500°C) at 2%	MCPA and fluometuron	Desorption	Reduced desorption in 85- to 3000-fold	Cabrera et al. [51]
Eucalyptus sp. (0.1–1%)	Isoproturon	Desorption	Reduced desorption in amended soils	Sopeña et al. [45]

Source: Adapted from Beesley et al. [57], Mesa and Spokas [23], and Khorram et al. [14].

Table 1.
Summary of recently selected studies using biochars and their influence on retention process of herbicides.

Another important feature affected by the pyrolysis temperature is the chemical composition of the biochar surface, so that the types of molecules present directly influence the retention of the herbicides. Sun et al. [44], analyzing the composition of wood and grass biochars under different thermal treatments (200–600°C), found the presence of cellulose, hemicellulose, and lignin groups (more present in wood biochar). At temperatures between 400 and 600°C, the noncarbonized aromatic carbon signals from the remaining lignin were identified and disappeared at temperatures above 500°C. The aliphatic carbon appeared between temperatures of 300 and 600°C, but above 600°C, was not pronounced. These variations interfere with the sorption capacity of the biochars, which in this experiment were exemplified by the sorption of fluridone in portions of aromatic biochar and values of sorption coefficient normalized as a function of the OC content (K_{oc}) higher than in the others, which were present in low temperatures.

Biochars that contain high specific surface area contribute to the increase of soil sorption of herbicides, as shown by Cabrera et al. [35], where they found almost complete sorption for aminocyclopyrachlor and bentazone in soils amended with biochar from wooden pallets (500°C). Sopeña et al. [45] indicated that biochar of *Eucalyptus dunni* with high specific surface area, sorption of isoproturon was five times higher than in unamended soils with biochar. When using wood of *Pinus radiata* as raw material for the production of biochar (350 and 700°C), Wang et al. [28] found increased terbuthylazine sorption in areas that had increased OC content of the soil and specific surface area with the addition of biochar. The total volume of pores in the soil indicates the greater or lesser specific surface area for the sorption of a herbicide. Biochar increases this amount of pores in the soil, as identified by Sandhu and Kumar [46]. The total pore volume is also related to the pyrolysis temperature, which, according to Wei et al. [47], is reduced at temperatures higher than 500°C for rice bark biochar, but there is increase in specific surface area (greater amount of micropores). In this same study, the pyrolyzed biochar at 750°C with a smaller pore diameter (9.23 nm) provided stronger sorption capacity for metolachlor due to intraparticle diffusion mechanisms and pore filling.

Besides the effect that the physical properties exert under the sorption of the different herbicides, the chemical properties also demonstrate strong influence in this process. The pH variation confers different sorption behaviors of ionic and nonionic molecules to the biochar as adsorbent in the soil. Sun et al. [44] observed that for norflurazon, the pH change had no effect on the sorption of this, because the nonionic molecules of the herbicide bind to the sorption sites and remain unaffected. For fluridone, sorption in both biochars (wood and grass) decreased with the pH increase because it ionizes at pH values lower than 4 (acid ionization constant—$pK_a = 1.7$); however, in regions in which the pH remained higher (pH 4–14), there was no ionization, and that with the increase of the proton concentration, there was the conversion of negative ionic functions to neutral sorption sites that fluridone probably binds.

Yang et al. [58] observed a decrease in sorption of diuron with increasing pH in the wheat biochar due to the alteration of the surface charge properties by deprotonation of the functional groups over the pH range. The study also evaluated pH variation in bromoxynil sorption, as well as Sheng et al. [59], both of which verified sorption of the herbicide in soil amended with wheat biochar was higher at low pH than at high pH. This fact can be justified because its pK_a is 4.06, and at $pH < 4.06$, this herbicide is in its molecular form, and already at $pH > 4.06$, it will be in the anionic form, which leads to the repulsion of loads on the soil and lower sorption of the product. In the case of ametryn, when in soil without biochar, the sorption of the herbicide occurred at lower pH and, when in addition to the biochar, the higher sorption occurred at higher pH, showing the increase of the affinity of the herbicide with the biochar due to protonation [60].

3.2 Desorption herbicides

The herbicide sorbed by the soil particles usually returns to the soil solution, that is, it is desorbed to be available again to the transport by absorption and degradation; however, this process in soils with biochar was less studied when compared to the sorption process [23].

In contrast to sorption, desorption decreases when biochar is added to the soil and the herbicide eventually returns to the soil solution, and sorption can often become irreversible. Irreversible sorption of herbicides was reported by Yu et al. [61], Wang et al. [28], and Sopeña et al. [45], which included sorption of the herbicides in the specific surface area, trapped in micropores, and partitioned into condensed structures of the biochar particles. In a study with wheat straw at 1% (m m^{-1}), Tatarková et al. [39] found the reduction of MCPA sorption from 64.2% in unamended soil to 55.1% in biochar amended soil. For Loganathan et al. [41], the amount of atrazine remaining in sandy and clayey soils amended with biochar was higher than in unamended soils. Cabrera et al. [35] found the almost negligible desorption for aminocyclopyrachlor in biochar amended soil from wood pallets (>500°C). Wang et al. [28] also reported slower and lower desorption rates for terbuthylazine in soil amended with sawdust biochar produced at 700°C, followed by biochar produced at 350°C. On the other hand, Khorram et al. [4] have reported easier desorption of fomesafen molecules weakly bound to rice bark biochar because the specific surface areas of the material are relatively low.

The reduction in herbicide desorption and consequently the lower concentration of these in the soil solution is more evident in altered soil with biochar in relation to the soil amended with the material. In this sense, there is, in general, a lower leaching potential of the herbicides and lower bioavailabilities of these, both for the degradation and the control of weeds.

4. Effect of biochar aged on herbicide sorption in soil

Another important factor in the properties of biochar is the action of time, which can alter the interaction of biochar with herbicides. Some physical processes such as breaking of the structure of biochar by the action of the climate can increase the specific surface area of the biochars [62] and the sorptive capacity of these materials. Kumari et al. [63] obtained higher specific surface area and cation exchange capacity (CEC) in soils amended with wood biochar (500°C) after 7–19 months, resulting in increased sorption of glyphosate in all soils of the study. Trigo et al. [64] observed an increase in the sorption of metolachlor in soils amended with distinct biochars over the years (macadamia: fresh = 2.4 times, 1 year = 2.5 times, 4 years = 1.9 times, wood: fresh = 2 times, and 5 years = 14 times). Martin et al. [38] evaluated sorption of atrazine and diuron in soil aged with biochar (10 t ha^{-1}) for 32 months. In this study, with fresh soil amended with biochar, there was an increase of twofold to fivefold in sorption of the herbicides in relation to the soil without biochar. With 5 years of aging, the biochar presented, in experiment with biochar of residues of the production of mushrooms and rice husk (70%) and peels of cotton seeds (30%) (400°C), increases in specific surface area of 98–114.3% according to Dong et al. [65]. However, the average pore diameter decreased and the surface was more propitious for material leaching. Structurally, there was no difference in fresh material.

On the other hand, in some studies, the aging of the biochar particles in the soil resulted in the reduction in the sorption capacity of the herbicides by reducing the specific surface area and the porosity of the biochar as the material aging, blocking the pores and the sorption sites, especially for high molecular weight molecules [66]. Martin et al. [66], in the same experiment mentioned previously, obtained for the diuron 47–68% reduction in the sorption capacity in relation to the control soil, which may be due to the clogging of the pores of the soil particles over time. Cao et al. [67], when aging rice husk biochar (500°C for 30 minutes) in soil, for a period of 13 months, found reduction in carbon and nitrogen content, reduction in pH values close to neutralized, and reduction in porosity and specific surface area of the biochar.

The degradation of the biochar particles throughout the weathering process alters their mass and their effect on herbicide remediation. Dong et al. [65] found a loss of biochar mass of ~40% over 5 years, regardless of the amount applied (30, 60, or 90 t ha^{-1}). Other authors have also observed loss of mass over time [68, 69]. Aging of the biochar can also entail structural changes in the material. Trigo et al. [54], analyzing the surface of aged biochars for 1 and 2 years, found clay minerals adhered by the addition of biochar to the soil, carboxylic acids covering the structure, and the elimination of fatty acids throughout the incubation periods, thus altering the types of possible connections to be made with this surface and the herbicide retention capacity. In the same study, after 2 years of incubation, the specific surface area on the biochar particles was even larger and the pores filled with mineral material. During the incubation period, there was a reduction in the OC content, from 80.4 mg L^{-1} of fresh biochar to 31.6 mg L^{-1} of biochar with 1 year of incubation, which may be due to natural elimination or degradation. However, these structural and chemical changes that alter the sorptive capacity of the biochar vary according to the nature of each material, as well as the pyrolysis temperature and the conditions and time at which the material will be incubated in the soil.

5. Conclusion

The application of biochar in the remediation of soil contaminant herbicides is an interesting management alternative due to high sorption and low desorption with these chemicals. Although the use is well explored worldwide, there is still a need for further research because of contradictory results regarding the benefits of using biochar and its effects on production, plant protection, and environmental contaminants. In addition to verifying the different interactions of the biochar and its properties in agricultural soils, it is worth emphasizing the importance of the possibility of using different materials in the production of biochars derived from different agricultural and industrial activities, which can promote the rational use of resources and destine them to agricultural production and minimization of the environmental impacts caused by these activities.

Acknowledgements

The authors would like to thank the São Paulo Research Foundation (FAPESP) process 2016/17683-1, for the financial support.

Author details

Kassio Ferreira Mendes[1*], Ananias Francisco Dias Júnior[2], Vanessa Takeshita[1], Ana Paula Justiniano Régo[1] and Valdemar Luiz Tornisielo[1]

1 Ecotoxicology Lab, Center of Nuclear Energy in Agriculture (CENA), University of São Paulo (USP), Piracicaba, SP, Brazil

2 Department of Forestry and Wood Sciences, Federal University of Espírito Santo (UFES), Jerônimo Monteiro, ES, Brazil

*Address all correspondence to: kassio_mendes_06@hotmail.com

References

[1] Lewis S, Silburn DM, Kookana R, Shaw M. Pesticide behavior, fate, and effects in the tropics: An overview of the current state of knowledge. Journal of Agricultural and Food Chemistry. 2016;**64**(20):3917-3924. DOI: 10.1021/acs.jafc.6b01320

[2] Yavari S, Malakahmad A, Sapari NB. Biochar efficiency in pesticides sorption as a function of production variables—A review. Environmental Science and Pollution Research. 2015;**22**(18):13824-13841. DOI: 10.1007/s11356-015-5114-2

[3] Jha P, Biswas AK, Lakaria BL, Rao AS. Biochar in agriculture—Prospects and related implications. Current Science. 2010;**99**(9):1218-1225

[4] Khorram MS, Wang Y, Jin X, Fang H, Yu Y. Reduced mobility of fomesafen through enhanced adsorption in biochar-amended soil. Environmental Toxicology and Chemistry. 2015;**34**(6):1258-1266. DOI: 10.1002/etc.2946

[5] Azargohar R, Dalai AK. Biochar as a precursor of activated carbon. Applied Biochemistry and Biotechnology. 2016;**131**(1-3):762-773. DOI: 10.1385/ABAB:131:1:762

[6] Sombroek W. Amazon Soils—A Reconnaissance of Soils of the Brazilian Amazon Region. Wageningen, Netherlands: Centre for Agricultural Publications and Documentation; 1966. 292 p

[7] Glaser B, Lehmann J, Zech W. Ameliorating physical and chemical properties of highly weathered soils in the tropics with charcoal—A review. Biology and Fertility of Soils. 2002;**35**(4):219-230. DOI: 10.1007/s00374-002-0466-4

[8] Maia CMBF, Madari BE, Novotny EH. Advances in biochar research in Brazil. Dynamic Soil, Dynamic Plant. 2011;**5**:53-58

[9] Mukherjee A, Lal R. The biochar dilemma. Soil Research. 2014;**52**(3):217-230. DOI: 10.1071/SR13359

[10] Yaman S. Pyrolysis of biomass to produce fuels and chemical feedstocks. Energy Conversion and Management. 2004;**45**(5):651-671. DOI: 10.1016/S0196-8904(03)00177-8

[11] Lehmann J, Rillig MC, Thies J, Masiello CA, Hockaday WC, Crowley D. Biochar effects on soil biota—A review. Soil Biology and Biochemistry. 2011;**43**(9):1812-1836. DOI: 10.1016/j.soilbio.2011.04.022

[12] Warnock DD, Lehmann J, Kuyper TW, Rillig MC. Mycorrhizal responses to biochar in soil-concepts and mechanisms. Plant and Soil. 2007;**300**(1-2):9-20. DOI: 10.1007/s11104-007-9391-5

[13] Laird DA, Koskinen WC. Triazine soil interactions. In: Lebaron HM, Macfarland JE, Burnside OC, editors. The Triazine Herbicides, 50 Years Revolutionizing Agriculture. San Diego, CA, USA: Elsevier; 2008. pp. 275-299

[14] Khorram MS, Zhang Q, Lin D, Zheng Y, Fang H, Yu Y. Biochar: A review of its impact on pesticide behavior in soil environments and its potential applications. Journal of Environmental Sciences. 2016;**44**:269-279. DOI: 10.1016/j.jes.2015.12.027

[15] Brown CD, Carter AD, Hollis JM. Soils and pesticide mobility. In: Roberts TR, Kearney PC, editors. Environmental Behaviour of Agrochemicals. Chichester, England: John Wiley & Sons; 1995. pp. 131-184

[16] Brito JO. O uso energético da madeira. Estudos Avançados.

2007;**21**(59):185-193. DOI: 10.1590/S0103-40142007000100015

[17] Gámiz B, Velarde P, Spokas KA, Hermosín MC, Cox L. Biochar soil additions affect herbicide fate: Importance of application timing and feedstock species. Journal of Agricultural and Food Chemistry. 2017;**65**(15):3109-3117. DOI: 10.1021/acs.jafc.7b00458

[18] Karunakara N, Kumara KS, Yashodhara I, Sahoo BK, Gaware JJ, Sapra BK, et al. Evaluation of radon adsorption characteristics of a coconut shell-based activated system for radon and thoron removal applications. Journal of Environmental Radioactivity. 2015;**142**:87-95. DOI: 10.1016/j.jenvrad.2014.12.017

[19] Andrade AD, Della Lucia RM. Avaliação da higroscopicidade do carvão vegetal e seus efeitos na resistência ao esmagamento. Floresta e Ambiente. 1995;**2**:20-26

[20] Yu XY, Ying GG, Kookana RS. Sorption and desorption behaviors of diuron in soils amended with charcoal. Journal of Agricultural and Food Chemistry. 2006;**54**(22):8545-8550. DOI: 10.1021/jf061354y

[21] Sharma RK, Wooten JB, Baliga VL, Lin X, Geoffrey Chan W, Hajaligol MR. Characterization of chars from pyrolysis of lignin. Fuel. 2004;**83**(11):1469-1482. DOI: 10.1016/j.fuel.2003.11.015

[22] Chen BL, Zhou DD, Zhu LZ. Transitional adsorption and partition of nonpolar and polar aromatic contaminants by biochars of pine needles with different pyrolytic temperatures. Environmental Science and Technology. 2008;**42**(14):5137-5143. DOI: 10.1021/es8002684

[23] Mesa C, Spokas KA. Impacts of biochar (black carbon) additions on the sorption and efficacy of herbicides. In: Kortekamp A, editor. Herbicides and Environment. Rijeka, Croatia: InTech; 2011. pp. 315-340

[24] Pétroff G, Doat J. Pyrolyse des bois tropicaux: Influence de la composition chimique des bois sur les produits de distillation. Revue Bois et Forêst Tropiques. 1978;**177**:51-64

[25] Somerville M, Jahanshahi S. The effect of temperature and compression during pyrolysis on the density of charcoal made from Australian eucalypt wood. Renewable Energy. 2015;**80**:471-478. DOI: 10.1016/j.renene.2015.02.013

[26] Clay SA, Malo DD. The influence of biochar production on herbicide sorption characteristics. In: Hasaneen MN, editor. Herbicides—Properties, Synthesis and Control of Weeds. Rijeka, Croatia: In Tech; 2012. pp. 59-74

[27] Yang XB, Ying GG, Peng PA, Wang L, Zhao JL, Zhang LJ, et al. Influence of biochars on plant uptake and dissipation of two pesticides in an agricultural soil. Journal of Agricultural Food and Chemistry. 2010;**58**(13):915-7921. DOI: 10.1021/jf1011352

[28] Wang HL, Lin KD, Hou ZN, Richardson B, Gan J. Sorption of the herbicide terbuthylazine in two New Zealand forest soils amended with biosolids and biochars. Journal of Soils and Sediments. 2010;**10**(2):283-289. DOI: 10.1007/s11368-009-0111-z

[29] Li J, Li Y, Wu M, Zhang Z, Lü J. Effectiveness of low-temperature biochar in controlling the release and leaching of herbicides in soil. Plant and Soil. 2013;**370**(1-2):333-344. DOI: 10.1007/s11104-013-1639-7

[30] Wang M, Zhu Y, Cheng L, Andserson B, Zhao X, Wang D, et al. Review on utilization of biochar for metal-contaminated soil and sediment remediation. Journal of Environmental

Sciences. 2018;**63**:156-173. DOI: 10.1016/j.jes.2017.08.004

[31] Gouveia SEM, Pessenda LCR, Aravena R. Dating of the humin fraction of soil organic matter and its comparison with ^{14}C ages of fossil charcoal. Química Nova. 1999;**22**(6):810-814. DOI: 10.1590/S0100-40421999000600007

[32] Piccolo A, Pietramellara G, Mbagwu JSC. Use of humic substances as soil conditioners to increase aggregate stability. Geoderma. 1997;**75**(3-4):267-277. DOI: 10.1016/S0016-7061(96)00092-4

[33] Gomes PA, Oliveira JB. Teoria da Carbonização da Madeira. In: Gomes PA, Oliveira JB, editors. Uso da Madeira para Fins Energéticos. Belo Horizonte: Fundo Centro Tecnológico de Minas Gerais/CETEC; 1982. pp. 27-42

[34] Singh B, Singh BPL, Cowie AL. Characterisation and evaluation of biochars for their application as a soil amendment. Soil Research. 2010;**48**(7):516-525. DOI: 10.1071/SR10058

[35] Cabrera A, Coxa L, Spokas K, Hermosín MC, Cornejo J, Koskinen WC. Influence of biochar amendments on the sorption–desorption of aminocyclopyrachlor, bentazone and pyraclostrobin pesticides to an agricultural soil. Science of the Total Environment. 2014;**470**:438-443. DOI: 10.1016/j.scitotenv.2013.09.080

[36] Kookana RS, Sarmah AK, Van Zwieten L, Krull E, Singh B. Biochar application to soil: Agronomic and environmental benefits and unintended consequences. Advances in Agronomy. 2011;**112**(12):103-143. DOI: 10.1016/B978-0-12-385538-1.00003-2

[37] Tang J, Zhu W, Kookana R, Katayama A. Characteristics of biochar and its application in remediation of contaminated soil. Journal of Bioscience and Bioengineering. 2013;**116**(6):653-659. DOI: 10.1016/j.jbiosc.2013.05.035

[38] Martin SM, Kookana RS, Van Zwieten L, Krull E. Marked changes in herbicide sorption–desorption upon ageing of biochars in soil. Journal of Hazardous Materials. 2012;**231**:70-78. DOI: 10.1016/j.jhazmat.2012.06.040

[39] Tatarková V, Hiller E, Vaculík M. Impact of wheat straw biochar addition to soil on the sorption, leaching, dissipation of the herbicide (4-chloro-2-methylphenoxy) acetic acid and the growth of sunflower (*Helianthus annuus* L.). Ecotoxicology and Environmental Safety. 2013;**92**:215-221. DOI: 10.1016/j.ecoenv.2013.02.005

[40] Loganathan VA, Feng Y, Sheng GD, Clement TP. Crop-residue derived char influences sorption, desorption and bioavailability of atrazine in soils. Soil Science Society of America Journal. 2009;**73**(3):967-974. DOI: 10.2136/sssaj2008.0208

[41] Xu C, Liu W, Sheng GD. Burned rice straw reduces the availability of clomazone to barnyardgrass. Science of the Total Environment. 2008;**392**(2-3):284-289. DOI: 10.1016/j.scitotenv.2007.11.033

[42] Spokas K, Koskinen WC, Baker JM, Reicosky DC. Impacts of woodchip biochar additions on greenhouse gas production and sorption/degradation of two herbicides in a Minnesota soil. Chemosphere. 2009;**77**(4):574-581. DOI: 10.1016/j.chemosphere.2009.06.053

[43] Hall KE, Spokas KA, Gámiz B, Cox L, Papiernik SK, Koskinen WC. Glyphosate sorption/desorption on biochars—Interactions of physical and chemical processes. Pest Management Science. 2018;**74**(5):1206-1212. DOI: 10.1002/ps.4530

[44] Sun K, Keiluweit M, Kleber M, Pan Z, Xing B. Sorption of fluorinated herbicides to plant biomass-derived biochars as a function of molecular structure. Bioresource Technology. 2011;**102**(21):9897-9903. DOI: 10.1016/j.biortech.2011.08.036

[45] Sopeña F, Semple K, Sohi S, Bending G. Assessing the chemical and biological accessibility of the herbicide isoproturon in soil amended with biochar. Chemosphere. 2012;**88**(1):77-83. DOI: 10.1016/j.chemosphere.2012.02.066

[46] Sandhu SS, Kumar S. Impact of three types of biochar on the hydrological properties of eroded and depositional landscape positions. Soil Science Society of America Journal. 2017;**81**(4):878-888. DOI: 10.2136/sssaj2016.07.0230

[47] Wei L, Huang Y, Li Y, Huang L, Mar NN, Huang Q, et al. Biochar characteristics produced from rice husks and their sorption properties for the acetanilide herbicide metolachlor. Environmental Science and Pollution Research. 2017;**24**(5):4552-4561. DOI: 10.1007/s11356-016-8192-x

[48] Yang Y, Sheng G. Enhanced pesticide sorption by soils containing particulate matter from crop residue burns. Environmental Science & Technology. 2003;**37**(16):3635-3639. DOI: 10.1021/es034006a

[49] Yang YN, Sheng GY, Huang MS. Bioavailability of diuron in soil containing wheat-straw-derived char. Science of the Total Environment. 2006;**354**(2-3):170-178. DOI: 10.1016/j.scitotenv.2005.01.026

[50] Cao XD, Ma L, Gao B, Harris W. Dairy-manure derived biochar effectively sorbs lead and atrazine. Environmental Science & Technology. 2009;**43**:3285-3291. DOI: 10.1021/es803092k

[51] Cabrera A, Cox L, Spokas K, Celis R, Hermosín MC, Cornejo J, et al. Comparative sorption and leaching study of the herbicides fluometuron and 4-chloro-2-methylphenoxyacetic acid (MCPA) in a soil amended with biochars and other sorbents. Journal of Agricultural and Food Chemistry. 2011;**59**(23):12550-12560. DOI: 10.1021/jf202713q

[52] Sun K, Gao B, Ro KS, Novak JM, Wang Z, Herbert S, et al. Assessment of herbicide sorption by biochars and organic matter associated with soil and sediment. Environmental Pollution. 2012;**163**:167-173. DOI: 10.1016/j.envpol.2011.12.015

[53] Dechene A, Rosendahl I, Laabs V, Amelung W. Sorption of polar herbicides and herbicide metabolites by biochar-amended soil. Chemosphere. 2014;**109**:180-186. DOI: 10.1016/j.chemosphere.2014.02.010

[54] Trigo C, Spokas KA, Cox L, Koskinen WC. Influence of soil biochar aging on sorption of the herbicides MCPA, nicosulfuron, terbuthylazine, indaziflam, and fluoroethyldiaminotriazine. Journal of Agricultural and Food Chemistry. 2014;**62**(45):10855-10860. DOI: 10.1021/jf5034398

[55] Eibisch N, Schroll R, Fuss R, Mikutta R, Helfrich M, Flessa H. Pyrochars and hydrochars differently alter the sorption of the herbicide isoproturon in an agricultural soil. Chemosphere. 2015;**119**:155-162. DOI: 10.1016/j.chemosphere.2014.05.059

[56] Hall KE, Ray C, Ki SJ, Spokas KA, Koskinen WC. Pesticide sorption and leaching potential on three Hawaiian soils. Journal of Environmental Management. 2015;**159**:227-234. DOI: 10.1016/j.jenvman.2015.04.046

[57] Beesley L, Moreno-Jiménez E, Gomez-Eyles JL, Harris E, Robinson

B, Sizmur T. A review of biochars' potential role in the remediation, revegetation and restoration of contaminated soils. Environmental Pollution. 2011;**159**(12):3269-3282. DOI: 10.1016/j.envpol.2011.07.023

[58] Yang Y, Chun Y, Sheng G, Huang M. pH-dependence of pesticide adsorption by wheat-residue-derived black carbon. Langmuir. 2004;**20**(16):6736-6741. DOI: 10.1021/la049363t

[59] Sheng G, Yang Y, Huang M, Yang K. Influence of pH on pesticide sorption by soil containing wheat residue-derived char. Environmental Pollution. 2005;**134**(3):457-463. DOI: 10.1016/j.envpol.2004.09.009

[60] Yamane VK, Green RE. Adsorption of ametryne and atrazine on an oxisoil, montmorillonite and charcoal in relation to pH and solubility effects. Soil Science Society of America Journal. 1972;**36**(1):58-64. DOI: 10.2136/sssaj1972.03615995003600010013x

[61] Yu XY, Pan LG, Ying GG, Kookana RS. Enhanced and irreversible sorption of pesticide pyrimethanil by soil amended with biochars. Journal of Environmental Sciences. 2010;**22**(4):615-620

[62] Lehmann J, Joseph S. Biochar for Environmental Management: Science, Technology and Implementation. Florence, Kentucky, USA: Routledge; 2015. p. 976

[63] Kumari KGID, Moldrup P, Paradelo M, Elsgaard L, Jonge LW. Soil properties control glyphosate sorption in soils amended with birch wood biochar. Water, Air & Soil Pollution. 2016;**227**(6):1-12. DOI: 10.1007/s11270-016-2867-2

[64] Trigo C, Spokas KA, Hall KE, Cox L, Koskinen WC. Metolachlor sorption and degradation in soil amended with fresh and aged biochars. Journal of Agricultural and Food Chemistry. 2016;**64**(16):3141-3149. DOI: 10.1021/acs.jafc.6b00246

[65] Dong X, Li G, Lin Q, Zhao X. Quantity and quality changes of biochar aged for 5 years in soil under field conditions. Catena. 2017;**159**:136-143. DOI: 10.1016/j.catena.2017.08.008

[66] Yavari S, Malakahmad A, Sapari NB. Biochar efficiency in pesticides sorption as a function of production variables—a review. Environmental Science and Pollution Research. 2015;**22**(18):13824-13841. DOI: 10.1007/s11356-015-5114-2

[67] Cao T, Chen W, Yang T, He T, Liu Z, Meng J. Surface characterization of aged biochar incubated in different types of soil. BioResources. 2017;**12**(3):6366-6377

[68] Maestrini B, Abiven S, Singh N, Bird J, Torn MS, Schmidt MWI. Carbon losses from pyrolysed and original wood in a forest soil under natural and increased N deposition. Biogeosciences. 2014;**11**(18):5199-5213. DOI: 10.5194/bg-11-5199-2014

[69] Obia A, Børresen T, Martinsen V, Cornelissen G, Mulder J. Vertical and lateral transport of biochar in light-textured tropical soils. Soil and Tillage Research. 2017;**165**:34-40. DOI: 0.1016/j.still.2016.07.016

Chapter 6

Impact of Physical/Chemical Properties of Volcanic Ash-Derived Soils on Mechanisms Involved during Sorption of Ionisable and Non-Ionisable Herbicides

Lizethly Caceres Jensen, Jorge Rodriguez Becerra and Mauricio Escudey

Abstract

Volcanic ash-derived soils (VADSs) are of great importance in the agricultural economy of several emerging and developing countries. The surface-charge amphoteric characteristics will confer physical/chemical properties absolutely different to constant-charge soils. This surface reactivity will confer to them a particular behaviour in relation to the herbicide sorption, representing an environmental substrate that may become polluted over time due to intensive agronomic uses. Sorption is a key parameter to evaluate the fate and behaviour of herbicides in volcanic soils. Sorption type and kinetic sorption models are also necessary in order to develop and validate QSAR models to predict pesticide sorption on volcanic soils to prevent potential contamination of water resources. The use of solute sorption mechanism models and QSAR models for pesticide sorption in soils has contributed to a better understanding of the behaviour of pesticides on volcanic soils. This chapter is divided into five sections: Physical/chemical properties of volcanic ash-derived soils; Ionisable and non-ionisable herbicides' fate and behaviour in soil; Kinetic sorption: mechanisms involved during sorption of ionisable and non-ionisable herbicides on VADS; Sorption of ionisable and non-ionisable herbicides on VADS; and Physical/chemical properties in QSAR models: a mechanistic interpretation.

Keywords: volcanic ash-derived soils, herbicides kinetic, herbicides sorption, solute transport mechanisms, mechanistic interpretation of QSAR models

1. Introduction

The nature of soils is regulated by various soil-forming factors such as parent material, climate, vegetation, relief and time [1]. These factors vary widely among region, and also vary in their properties. Volcanic ash-derived soils (VADSs) are predominantly found in regions of the world with geochemical characteristics dominated by

IntechOpen

active and recently extinct volcanoes. These have great importance in the agricultural economy of several emerging and developing countries of Europe, Asia, Africa, Oceania and America. They are abundant and widespread in Central-Southern Chile (from 19° to 56° S latitude), accounting for approximately 69% of the arable land [2].

Agricultural practices developed in Chilean VADS (ChVADS) have led to the very increased use of pesticides and also frequent adjustments of soil pH and mineral fertilisation [3–5]. Among these soils, andisols and ultisols are the most abundant and present an acidic pH (4.5–5.5). Andisols are characterised by their high organic carbon (OC) content, highspecific surface area and a mineralogy dominated by short-range-ordered minerals such as allophane ($Al_2O_3SiO_2 \times nH_2O$). Ultisols have lower OC than andisols, but higher total iron oxide content. Andisols present variable surface charge, originated in both inorganic and organic constituents. Inorganic minerals as goethite (FeOOH), ferrihydrite ($Fe_{10}O_{15} \times 9H_2O$), gibbsite ($Al(OH)_3$), imogolite and allophane contribute through the dissociation of Fe-OH and Al-OH-active surface groups; while organic mineral (OM) contributes through the dissociation of its functional groups (mainly carboxylic and phenolic), and humus-Al and Fe complexes with amphoteric characteristics. Nevertheless, ultisols present lower variable surface charge than andisols, because more crystalline minerals such as halloysite and/or kaolinite dominate their mineralogy.

Several sorption kinetic studies of herbicides on VADS have indicated that herbicide sorption is a non-equilibrium process [5]. Time-dependent sorption (or non-ideal sorption) can be a result of physical and chemical non-equilibrium and intra-sorbent diffusion that can occur during the transport of pesticides in soils [6, 7]. In general, non-equilibrium sorption has been attributed to several factors such as diffusive mass transport resistances, non-linearity in sorption isotherms, sorption-desorption non-singularity and rate-limited sorption reactions [8]. Intra-OM diffusion has been suggested to be the predominant factor responsible for the non-equilibrium sorption of non-ionic or hydrophobic compounds on VADS [7, 9]. It has been found that differences in sorption kinetic of herbicides were due to soil constituents, such as OC and mineral composition on VADS.

In general, sorption processes are known to be important because they are time dependent and with considerable ecosystem impact, influencing the availability of organic pollutants for plant uptake, microbial degradation and transport in soil and, consequently, leaching potential. In this sense, the principal process that affects the fate of pesticides in soil and water is the sorption of pesticides from soil solution to soil particle active sites, which limit transport in soils by reducing their concentration in the soil solution.

The kinetic parameters can be obtained by means of the application of two kinds of kinetic models: the ones that allow to establish principally kinetic parameters and modelling of the sorption process and other models frequently used to describe sorption mechanisms of organic compounds on soils. Such information is necessary in order to understand leaching of pesticides, such as herbicides for preventing potential contamination of groundwater.

The aim of this chapter is to establish the sorption kinetics of ionisable and non-ionisable herbicides (INIH) in ChVADS to apply different solute sorption mechanism models, considering the models' restrictions and VADS properties to investigate the mechanisms involved in INIH sorption on VADS. These kinetics studies, complemented with 'batch' sorption studies of INIH on VADS, allow the identification of sorption characteristics. Sorption type and kinetic sorption models description are also necessary in order to develop and validate computer simulation transport models on VADS or to increase the quality of sorption data to develop reliable models, such as QSAR models, and to predict pesticide sorption on VADS to prevent potential contamination of water resources.

2. Physical/chemical properties of volcanic ash-derived soils

In general, VADSs are soils rich in constituents with amphoteric surface reactive groups; Although andisols and ultisols are the most important in Chile, oxisols, alfisols and spodosols are also considered variable charge soils [10]. The nature of soils is regulated by various soil-forming factors such as parent material, climate, vegetation, relief and time [1]. These factors vary widely among regions, also affecting their properties. The most striking and unique properties of these are variable charge, high water-holding capacity, low bulk density, high friability, highly stable soil aggregates, excellent tilth and strong resistance to water erosion [11], high anion sorption, high lime or gypsum requirement to achieve neutral pH and considerable sorption affinity for cations (Ca and Mg), which may form both inner- and outer-sphere surface complexes although the first is found to be most important [10].

These distinctive physical and chemical properties are largely due to the presence of non-crystalline materials, biological activity and the accumulation of OC [11, 12]. The soil organic matter (SOM) represents a key indicator of soil quality, both for agricultural (i.e. productivity and economic returns) and environmental functions (i.e. carbon sequestration). The OC concentrations in andisols are more strongly associated with metal-humus complexes than with concentrations of non-crystalline materials; nevertheless, inorganic materials with variable-charge surfaces provide an abundance of microaggregates that permit to encapsulate OC, favouring their physical protection [11]. Other studies indicate Al/Fe oxides/hydroxides in allophanic soils are linked through carboxylic and aromatic groups present in SOM being the SOM highly decomposed [1].

VADSs are dominated by Al/Fe-humus complexes, by ferrihydrite, a short-range-order Fe hydroxide mineral or by short-range-order clay components (amorphous aluminosilicates), such as allophane and imogolite [11]. The VADS clay fraction mineralogy is usually dominated by kaolinite, gibbsite, goethite and hematite [10]. Besides these minerals, they contain 2:1 and 2:1:1-type minerals and opaline silica, halloysite, etc. occasionally in substantial or dominant amounts. Halloysite is a 1:1 aluminosilicate hydrated mineral characterised by a diversity of morphologies (e.g. spheroidal and tubular) [11].

Andisols are relatively young soils and cover about 0.84% of the world's land [11, 13] being typical products of weathering increases in both temperate and tropical environments with sufficient moisture [11]. In this sense, metastable non-crystalline materials are transformed to more stable crystalline minerals (e.g. halloysite, kaolinite and gibbsite) allowing the alteration of andisols to inceptisols, alfisols or ultisols. Andisols are often divided into two groups based on the mineralogical composition of A horizons, with allophanic andisols dominated by variable-charge constituents (allophane/imogolite), and non-allophanic andisols dominated by both variable-charge and constant-charge components (Al/Fe-humus complexes and 2:1 layer silicates) [11]. Allophanic andisols form preferentially in weathering environments with pH values in the range of 5–7 and a low content of complexing organic compounds. Non-allophanic andisols form preferentially in pedogenic environments that are rich in OM and have pH values of 5 or less [11].

Allophanic andisols of Southern Chile derive from holocenic volcanic ash, presenting dates less than 12,000 years old. Chilean andisols are rich in OM, with high specific surface area and a mineralogy dominated by short-range-ordered (amorphous) minerals such as allophane, high P retention (>85%), low saturation of bases, presence of clay, high variable charge, low bulk density (<0.9 Mg m^{-3}) associated with a high porosity and a strong microaggregation of heterogeneous forms [2, 14]. The variable surface charge in Chilean andisols is originated in both organic and inorganic constituents. The OM contributes through the dissociation of its functional groups (mainly carboxylic and phenolic) and Al/Fe-humus complexes with amphoteric characteristics; while inorganic minerals such as goethite, ferrihydrite, gibbsite, imogolite and allophane contribute through the dissociation of Si-OH, Fe-OH and Al-OH active surface groups [15]. Furthermore, allophane plays key roles in surface reactivity such as determining the availability of nutrients and controlling soil contaminant behaviour [14].

2.1 Agricultural implications of Chilean volcanic ash-derived soils

The importance of VADSs is due to the ability to manipulate their surface charge characteristics in order to control the retention of cations and anions[10]. Chilean andisols have higher total P concentrations than ultisols, and significant amounts of the accumulated P are from the organic forms (organic P) (>45% of the total P), similar to allophanic soils in other parts of the world [4, 5]. The inorganic P fraction has been associated with Fe and Al in uncultivated Chilean andisols, and organic P has been strongly correlated with OC content, being considered an important P source for crops through mineralisation. But generally, they have low available P resulting in reduced fertility [2, 4, 5, 16, 17]. The availability of P decreases or increases in relation to the development of the soil, the decrease of P availability with increasing soil development due to incorporation of P into organic forms and P fixation by non-crystalline Al and active Al/Fe components making it sparingly available for plant uptake [12]. On the other hand, P is often a growth-limiting nutrient for agricultural crops grown on relatively young soils where its availability is relatively high due to rapid weathering of apatite, and its retention is low due to low concentrations of active Al/Fe [11].

At their original acidic pH range (4.5–5.5), VADSs require frequent adjustments of soil pH, replenishment of exchangeable Mg and P applications to remain productive [4]. In relation to adjustments of pH, the increase of soil pH has been shown to reduce phytotoxic levels of exchangeable acidity (Al^{3+}) and increase the ability of soils to retain nutrient ions and potential toxic heavy metals [10]. The exchangeable Al often dominates exchange sites, controls soil acidity and buffering capacity on ultisols, resulting in heavy reliance on liming practices to optimise soil acidity/fertility for plant growth [18].

Although P fertiliser application has been proposed as a management tool to increase the Cation-exchange capacity (CEC) of volcanic charge soils (VCS), large quantities of P fertiliser are required to cause a significant increase in CEC [10]. P fertiliser applied to andisols is rapidly sorbed by active Al/Fe components and changes to less available forms with time [11]. In general, ChVADS presents a high capacity to retain P due to its specific phosphate sorption [2]. Also, specific and preferential sorption of phosphate by variable-charge minerals can modify the soil surface to be more negative [17]. In such cases, it could result in the enhanced mobility of ionisable herbicides, because of increased competition with inorganic anions for positively charged sites and could modify the charge on the oxide surface, changing their speciation. Moreover, the strong influence of pH and phosphate addition in its sorption on ChVADS are conditions that are known to favour excessive transport of sulfonylurea herbicides (SHs), such as metsulfuron-methyl (MSM) [5]. A displacement of the isoelectric point (IEP) of an allophanic andisol towards a higher pH would be a most favourable condition for electrostatic interaction between anionic MSM and free or active Fe/Al oxides. The SH anionic forms predominate in solution for most soils, which is more soluble in water, being less susceptible to hydrolysis but favouring the herbicide transport. This situation is aggravated when considering that ChVADSs are relatively shallow (<15 m) in relation to groundwater.

Therefore, the lower sorption of MSM at higher pH was attributed mainly to the decrease of available active sites, despite its high maximum sorption capacity of phosphate. Maximum phosphate sorption capacity for ultisols (mainly attributed to kaolinite content) was 1.7 times lower than for andisols, so a lower amount of common sites will be also available for MSM sorption. In this sense, intensive soil fertilisation and liming are the most probable scenarios for leaching potential of ionisable herbicides in VADS as a consequence of decreasing soil sorption. The extensive use of glyphosate (GPS) on Chilean andisols may result in enhanced sorption and, consequently, reduced availability of phosphate [4]. Moreover, on Chilean ultisols, GPS may be immobilised with no effect on the phosphate sorption, suggesting a different mechanism involved during GPS and phosphate sorption on VADS. In this sense, the long-term use of GPS may therefore have different effects on the retention and availability of soil P. The continuous input of P fertiliser for VADS may subsequently decrease sorption of post-applied carboxylic acid herbicides, such as 2,4-dichlorophenoxyacetic acid (2,4-D) and desorb previously applied 2,4-D [19]. In this sense, Chilean andisols are generally cropped to maize (*Zea mays* L.) and their management includes the application of liquid cow manure (LCM) at rates higher than 100,000 L ha^{-1} and the application of atrazine for broad leaf weed control [14].

3. Ionisable and non-ionisable herbicides fate and behaviour in soil

Herbicides are the dominant pesticides used to control weeds in agricultural production. The total amount of pesticides used in the world exceeded 39.4 billion USD in 2007, of these, herbicides accounted for the largest amount (40%) [20]. The physical, chemical and biological characteristics of soil, as well as the chemical properties of herbicides, will influence their fate and behaviour in soils [1, 19, 21, 22]. And also, sorption processes are an important physicochemical characteristic on use herbicides, which are used considering their specific biological activity on target species.

Even more, although sorption-desorption processes are dynamic, where molecules are continually transferred between the bulk liquid and solid surface, an excess of herbicide sorption may result in unavailability of herbicide to targeted pests as well as uneven distribution around the plants. In this sense, sorption is a key parameter to evaluate the fate and behaviour of herbicides in soils, controlling the bioavailability, distribution and transport to other environmental compartments. In this regard, the main processes of herbicides in soils, such as sorption, degradation, biodegradation, bioavailability and transport, are commonly studied and evaluated [22].

Soil sorption is characterised by a partition coefficient, K, conventionally written with a subscript d ('distribution'). The distribution coefficient (K_d) is the most common and accepted quantitative measurement of pesticide soil sorption [23, 24]. Moreover, it is considered as a unique property or constant of pesticides, which is used to describe the equilibrium distribution of a pesticide between a soil, sediment or particles and the aqueous phase that it is in contact with [25]. Several mathematical models have been developed to describe equilibrium sorption of pesticides on soils, such as *Linear*, *Freundlich* and *Langmuir* models. A *linear* model assumes that the relationship between the amount adsorbed and concentration at equilibrium of the chemicals is proportional which is often only valid at trace levels, for example, below half solubility. The *Freundlich* model assumes that the solid matrix has an infinite sorption capacity, thus sorption increases indefinitely with the solute concentrations in a non-linear way [26]. At low pesticide concentration levels, this model has been widely applied describing adequately the sorption behaviour of

INIH on VADS (**Figure 1**). *Freundlich* model with (1/n) < 1 (L-type) indicates a heterogeneous sorption site, a strong adsorbent affinity for the adsorbate, diversity of sorption mechanisms and strong concentration dependence of sorption for the sorption sites [4, 14, 15, 19]. Isotherms with (1/n) > 1 (S-type) indicate competition between solvation water and adsorbate for the sorption sites [19]. The *Langmuir* model assumes: (1) sorption occurs on planar surfaces that have a fixed number of sites that are identical and the sites can hold only one molecule; thus, only monolayer coverage is permitted, which represents maximum sorption; (2) sorption is reversible; (3) there is no lateral movement of molecules on the surface and (4) the sorption energy is the same for all sites and independent of surface coverage (i.e. the surface is homogeneous), and there is no interaction between adsorbate molecules (i.e. the adsorbate behaves ideally) [26].

4. Kinetic sorption: mechanisms involved during sorption of ionisable and non-ionisable herbicides on VADS

Sorption processes are known to be important because they are time-dependent processes with considerable ecosystem impact, influencing the availability of organic pollutants for plant uptake, microbial degradation and transport in soil and consequently the leaching potential. Sorption kinetic studies may provide important information related to weed control, crop toxicity, runoff, sorption mechanisms, solute transport mechanisms and its applied use in decontamination, such us remediation [7, 22]. This process occurs within the boundary layer around the sorbent, being conceptualised as a rapid uptake process to readily available sorption sites and proceeds in the liquid-filled pores (external mass transfer steps; (EMT)) or along the walls of the pores of the sorbent (internal mass transfer steps (IMT)) followed by slow diffusion-immobilisation in micropores or capillaries of the sorbent's internal structure (intra-particle diffusion, IPD), except for small sorbed amounts that appear on the external surface, and can occur during the transport of pesticides in soils. The third stage is the sorption of the solute in the inner surface of the sorbent through mass-action-controlled mechanisms where a rapid uptake occurs or surface reaction through interactions between solute and surface functional groups (such as chemisorption) [7, 9, 27].

The herbicide-VADS interaction is a time-dependent process that often progresses rapidly over the short term (minutes or hours), and it may also take a short time to reach equilibrium on VADS (**Figure 1**). Sorption process has influence on the transport of pesticides in the soil environment during the short term [4, 7, 9, 15]. The *pseudo-second-order* model has been the best sorption kinetic model to establish principal kinetic parameters and modelling of sorption process of INIH on VADS (**Figure 1**). On the other hand, the *Weber-Morris* model is one of the most used models to describe solute transport mechanisms of organic compounds in different sorbents intended for remediation purposes. Nevertheless, the *two-site non-equilibrium* (TSNE) model has been the best kinetic model to describe INIH transport mechanisms on VADS (**Figure 1B**; **Table 1**).

Figure 1A and **B** shows different transport mechanisms of MSM and diuron (DI) on ultisol (Collipulli soil, COLL) and andisol (Nueva Braunau soil, NBR). The *Weber-Morris* model indicates that the mass transfer across the boundary layer and IPD control DI sorption on all ChVADSs (**Figure 1A**; **Table 1**). The non-ionic or hydrophobic herbicide sorption on VADS has been described as a two-site equilibrium-kinetic process, where intra-OM diffusion has been suggested to be the predominant factor responsible for non-equilibrium sorption on andisols. In contrast, the MSM sorption on ultisols was controlled exclusively by IPD; thus, the

first line (at shorter time) will depict macropore diffusion in the boundary layer, with the EMT occurring at short contact times during the retention of MSM into macropores and the second line (at longer time) accounts for the gradual sorption

Pesticide-VADS	**Kinetics sorption description and/or model**	**Ref.**
AT-Andisol (Southern, Chile)	AT approached equilibrium in about 12 h	[14]
AT-acidic andisols and ultisols (Southern Chile). Soils presented negative net charge at pH_{soil}	AT sorption on VADS was controlled by instantaneous equilibrium followed by a time-dependent phase. Andisols: two rate-limited phases were established; the first one related to intra-sorbent diffusion in OM and second one related to slow IPD in the organo-mineral complex. Ultisols: only one rate-limited phase attributed to kaolinite associated to slow and progressive sorption	[15]
GPS-acidic ultisol and andisol (Southern Chile)	GPS sorption kinetics on ChVADS followed a *pseudo-second-order* model with an apparent equilibrium reached in 10–120 min. The faster GPS sorption of andisols was related to higher OC content. The slower GPS sorption kinetic on ultisols was controlled by IPD	[4]
T-non-allophanic (Central Chile) and allophanic soils (Southern Chile). Soil pH range: 6.3–7.4	A rapid adsorption was observed during the first few minutes, followed by a slower process that in all cases reached an apparent equilibrium within 2 h. The order of the kinetic reaction was two. The slow process may be attributed to the diffusion of MBT within the porous of the soil matrix. On the sorption kinetic for non-allophanic soils, OM amendment increased the X_{max} values and decreased the rate constant, pointing out a higher porosity of the soil matrix	[33]
MSM-acidic ultisols and andisols (Southern Chile)	MSM sorption kinetics on ChVADS followed a *pseudo-second-order* and the *Weber-Morris* model, indicating that the mass transfer across the boundary layer and IPD are the two mechanisms controlling MSM sorption on andisol, whereas in ultisols, the rate was controlled exclusively by IPD into macropores/micropores. The *two-site nonequilibrium* (TSNE) model was the best kinetic model to be applied to VADS. Andisol presented an initial phase with a fast trend to equilibrium, where ~50% of sites accounted for instantaneous MSM sorption. Most ultisol sites corresponded to the time-dependent stage of sorption	[9]
DI-acidic ultisols, inceptisol and andisol (Southern Chile)	The *pseudo-second-order* model was able to describe DI sorption at all time intervals for all soils. The *Weber-Morris* model indicated that mass transfer across the boundary layer and IPD were the two processes controlling sorption kinetics of DI in all ChVADSs, being corroborated by the *Boyd* model. An initial phase, with a fast trend to equilibrium, was established for the andisols through the TSNE model, where ~ 51% of sites account for instantaneous sorption on andisols. For the ultisols, most of the sites corresponded to the time-dependent stage of sorption	[7]
Pesticide-VADS	Sorption description model	Ref.
AT-Andisol (Southern Chile)	AT sorption on ChVADS was well described by the *Freundlich* model (K_f = 1.19 $mg^{(1-N)}L^{N}kg^{-1}$; R^2 = 0.98), exhibiting non-linear isotherms of L-type ($N_f < 1$) and a concentration-dependent AT solid-solution distribution. The mechanisms involved are hydrogen bonding and charge transfer. AT in acidic soil, for its weak basic nature, presents a partial protonation of amine groups, which are involved in the hydrogen bonds with carbonyl and carboxyl groups of soil. These contribute to its protonation, forming a partial positive charge in the aromatic ring	[14]
AT-acidic andisols and ultisols (Southern Chile). Soils presented negative net charge at pH_{soil}	Sorption data were well described by the *Freundlich* model (K_f between 2.2 and 15.6 $\mu g^{1-1/n}mL^{1/n}g^{-1}$; $R^2 \geq 0.993$). The highest sorption was observed on andisols being adsorbed mainly through hydrophobic interactions or van der Waals forces at the phenolic groups of OM (humic acid and humin fraction). AT was weakly sorbed on soil with permanent negative charge (ultisols) through a hydrophobic bonding at the silanol surface sites of the tetrahedral sheets present on clays, such as chlorite, gibbsite, goethite/hematite and kaolinite	[15]

Pesticide-VADS	Kinetics sorption description and/or model	Ref.
BSM-acidic andisols and ultisols (Southern Chile) which presented negative net charge at pH_{soil}	BSM sorption on ChVADS was well described by the *Freundlich* model (R^2 = 0.995), exhibiting non-linear isotherms of L-type. BSM was weakly sorbed on soil with permanent negative charge (ultisols). A multifactorial influence of several properties of soils was demonstrated, through a partial least squares regression model (P value = 0.0042) by use of % OC, % Fe and surface area. The BSM sorption on mineral (AlSi-Fe) and mineral-organic complexes (AlSi-Fe-HA) confirmed the participation of variable-charge materials present in VADS, both with a high sorption capacity, demonstrating the effect of BSM sorption on charge of both sorbents. The distribution of SOH, SOH_2^+ and SO^- sites was better characterised for AlSiFe-HA	[16]
GPS-acidic ultisol and andisol (Southern Chile)	GPS sorption on ChVADS over the extended range of concentrations was well described by *Freundlich* model (K_f between 1480 and 3764 $\mu g^{1-1/n}$ $mL^{1/n}g^{-1}$; $R^2 \geq 0.99$). The highest sorption was observed on andisols being adsorbed to Fe/Al oxides and allophane by ligand exchange through its phosphonic acid moiety or by hydrogen bonding reacting with polyvalent cations adsorbed on SOM between GPS and humic substances (metal-GPS OM complex). GPS was adsorbed strongly and specifically to kaolinite on ultisol	[4]
AT, CPF, CTL, DZN, GPS and MSM, DEA, AMPA, TCP and IMHP-acidic ultisol and andisol (Southern Chile). All soils present a negative net charge at pH_{soil}	Data fitted to the *Freundlich* model with $R^2 > 0.97$. All herbicides and metabolites (except GPS) showed a markedly higher sorption on allophanic soil exhibiting a non-linear isotherm of L-type ($N_f < 1$) with a trend to the saturation of sorption sites at higher concentrations. The exceptionally high sorption of GPS on ultisols (with lower OM, Al_{Ox} and Fe_{Ox} contents than andisols; K_f>1500 $\mu g^{1-1/n}mL^{1/n}g^{-1}$) was related to kaolinite contents and acidic pH, posing the need to establish its possible contamination	[3]
MBT-non-allophanic (Central Chile) and allophanic (Southern Chile) soils. The pH values of natural soils (6.3–7.4)	The isotherms fit the *Freundlich* model (K_f between 5.3 and 82.1 $cm^3\ g^{-1}$; $R^2 \geq 0.998$) exhibiting a non-linear isotherm of L-type ($N_f < 1$). The MBT sorption was markedly higher on allophanic soils with OM being the principal component responsible for MBT sorption. The clay and smectite contents on non-allophanic soils were considered the most important factors governing the MBT sorption. The interaction was predominantly by physical bonding (van der Walls forces or hydrogen bonding) between the amino hydrogen of MBT and hydroxyl groups of humic acids of SOM, or through the carbonyl group, which could act as a strong donor to hydrogens of the alcoholic and phenolic groups on the humic acid of SOM. The presence of MBT in surface waters in Chile would be produced only by losses from the sediment-adsorbed MBT through runoff process	[33]
MSM-ultisol and andisol (Southern Chile)	The *Freundlich* model described MSM sorption in all ChVADSs (K_f values between 3.1 and 14.4 $\mu g^{1-1/n}mL^{1/n}g^{-1}$; $R^2 > 0.992$). The lower MSM sorption capacity on low variable-charge soils (Chilean ultisols) was attributed to the low OM content. The kaolinite mineral group as major constituent of the inorganic fraction of ultisols and minerals, such as allophane, gibbsite, hematite and goethite, contributed to MSM sorption mainly through hydrophilic interactions. The OM and active/free Fe/Al oxides controlled the MSM sorption in andisols mainly through hydrophilic rather than hydrophobic interactions	[5]
2,4-D-oxisols (Brazil), andisols (South Korea), ultisols (Costa Rica), and alfisols (Toronto)	Linear sorption model fitted to isotherms measured from $CaCl_2$, $CaSO_4$, $Ca(H_2PO_4)_2$ and KCl systems (K_d values between 1.28 and 18.5 L Kg^{-1}; $R^2 > 0.96$). The higher 2,4-D sorption was obtained on $CaCl_2$, reflecting both effects of inorganic anion in terms of competition and cation in terms of bridging interactions. 2,4-D sorption via Ca-bridging took place on silanol edges (Si-OH) of kaolinite or silicate edges	[17]

Pesticide-VADS	Kinetics sorption description and/or model	Ref.
2,4-D-SOM-rich acidic andisol (Japan)	Surface area, Al/Fe oxide content, OC content, pH, soil phosphate and exchangeable Al content, and active surface hydroxyls derived from the active and free metal (hydr)oxides, such as allophane, imogolite, ferrihydrite, goethite and metal-SOM complexes might have an important role in the carboxylic acid herbicide sorption. 2,4-D sorption on andosol was regulated by ion exchange reaction and/or a ligand exchange reaction in which the active surface hydroxyls on Al and Fe were replaced by the carboxylic group of 2,4-D. 2,4-D may form surface complexes with exchangeable Al ions (via a cation-bridging mechanism involving an exchangeable Al and the carboxylate group of 2,4-D) or be electrostatically attracted to the positively charged exchangeable Al ion	[31]

Atrazine (AT), bensulfuon-methyl (BSM), chlorpyrifos (CPF), chlorothalonil (CTL), deethylatrazine (DEA), diazinon (DZN), diuron (DI); glyphosate (GPS) and methabenzthiazuron (MBT), metsulfuron-methyl (MSM), aminomethylphosphonic acid (AMPA), 3,5,6-trichloro-2-pyridinol (TCP); isopropyl-4-methyl-6 hydroxypyrimidine (IMHP) and 2,4-dichlorophenoxyacetic acid (2,4-D).

Table 1.
Kinetics sorption and sorption-desorption of INIH on VADS.

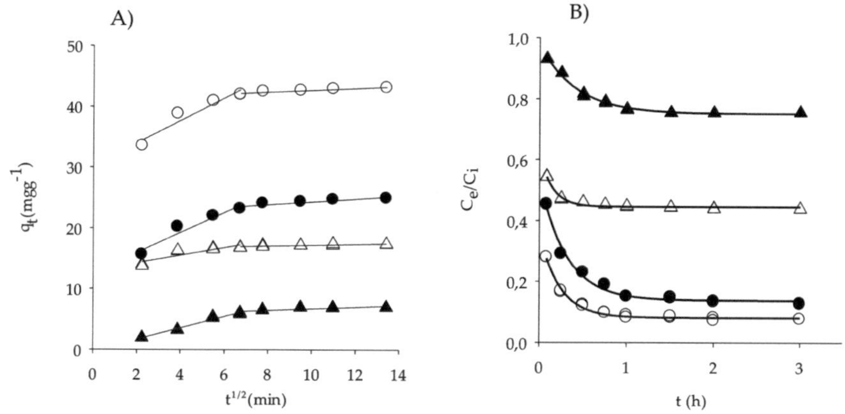

Figure 1.
(A) IPD plots for MSM kinetic sorption on ultisol (▲) and andisol (Δ); and DI kinetic sorption on ultisol (•) and ndisol (O); (B) TSNE model plot for MSM sorption on ultisol (▲) and andisol (Δ) and DI sorption on ultisol (•) and andisol (O) [7, 9].

stage, where the molecules of MSM diffuse through the smaller pores of the soil (IPD) (**Figure 1A**; **Table 1**). On the other hand, the mass transfer across the boundary layer and IPD were the two mechanisms to control MSM sorption on andisol (**Figure 1A**; **Table 1**).

Time-dependent sorption (or non-ideal sorption) can be a result of physical and chemical non-equilibrium [9]. Non-equilibrium sorption on soils has been attributed to several factors, such as: diffusive mass transfer resistances, non-linearity in sorption isotherms, sorption-desorption non-singularity and rate-limited sorption reactions [8]. The rate-limited diffusion of the sorbate from bulk solution to the external surface of the sorbent, and rate-limited diffusion within mesopores and micropores of the soil matrix, will occur before the equilibrium is reached. The TSNE model (**Figure 1B**; **Table 1**) indicated that MSM sorption on andisol presented an initial phase with a fast trend to equilibrium, where about 50% of sites accounted for the instantaneous stage and the great part of sites on ultisols corresponded to the time-dependent stage of sorption (90%). In contrast, the sorption of non-ionisable herbicide (DI) on andisols presented an initial phase, with a fast trend to equilibrium, where~20% of sites accounted for instantaneous sorption on andisol. For the ultisols, most of the sites corresponded to the time-dependent stage of sorption (90%).

The higher value in the overall rate constant, k_2, of MSM on andisols with respect to DI in the same soil indicates that this value reflects contributions from the favoured electrostatic interactions considering both a retarded IPD as well as intra-OM diffusion. The way minerals present on VADS are interrelated or chemically spatially distributed, either being freely distributed throughout the soil mass or coating silt and clay grains, is determinant of their chemical role in the whole ion sorption-desorption mechanisms [28]. In this sense, the OM content is the principal component to control the pesticide sorption on andisols, as much by instantaneous equilibrium as by IPD, the presence of kaolinite, halloysite and Al/Fe oxides in Ultisols will be significant in the IPD mechanism. According to this analysis, ultisols present a potential risk of ionisable herbicide transport. The different mineral composition of ultisols impacts on their different physical behaviour, influencing the slowest INIH sorption rate, the sorption mechanism involved and the lowest INIH sorption capacity. All of the above must be taken into account to evaluate the potential leaching of INIH in these kinds of soils.

5. Sorption of ionisable and non-ionisable herbicides on VADS

The retention and mobility of herbicides in soil are determined by the strength and extent of sorption reactions [14]. The soil particles (adsorbent) may adsorb herbicides (adsorbate) weakly or strongly depending on the adsorbent-adsorbate interactions. In this sense, herbicides can be adsorbed in soil through different mechanisms such as physical sorption (van der Waals and H bond interactions) and chemical sorption. Physical sorption is fast and usually reversible, due to small energy requirements [21].

The physical and chemical interactions between herbicides and other organic molecules on the soil particles surface depend on the physical and chemical properties of the soil and herbicide. The nature of surface charge may vary with the chemical properties of VADS. The surface charge amphoteric characteristics will confer to VADS chemical and physical properties absolutely different from those exhibited in soils of constant charge. This surface reactivity of VADS confers to them a particular behaviour in relation to the retention of herbicides, representing an environmental substrate that may become polluted over time due to intensive agronomic uses.

There are only a few reports on the behaviour of INIH in VADS despite being important to agricultural systems of many regions (**Figure 1**). A higher sorption capacity of several herbicides has been reported for allophanic soils (**Figure 1**). In this regard, the herbicide sorption on VADS will be affected by soil properties (SOM content, allophane, clay, pH, IS, particle size distribution, moisture content and variable charge) and herbicide chemical properties (molecular structure, molecular size, electrical charge, ionisability, aqueous solubility, hydrophobicity (K_{ow}), volatility, reactivity with soil constituents and longevity in the environment) [29]. Environmental conditions may also affect INIH sorption and mobility on VADS.

5.1 Sorption of ionisable herbicides and non-ionisable herbicides on VADS

The ion sorption rate in VADS depends on the surface area, CEC, the proportion of Fe/Al oxides or oxyhydroxides present as the surface coating of clay and silt particles [28]. Oxides have also been found to enhance the deprotonation of organic acids and, therefore, increase the activity of the anionic species at a given is expected to be at a maximum if the ratio of the mineral to the OC fractions is more than 30, regardless of the mineral content [7, 9, 19].

At soil pH, both Fe/Al oxides and amorphous Fe oxide surfaces on VADS are positively charged. This property is considered effective in the retention of different SH at pH_{soil} > ZPC, oxides and amorphous Fe oxide surfaces on VADS has been considered the most effective condition for the retention of different SH at pH_{soil} > zero point of charge (ZPC), where the anionic herbicide will be present (**Figure 1**), due to the strong dependence of pH in sorption of this kind of compound. Caceres et al. [5] studied the effect of the sorption of MSM on VADS. These researchers observed a change in the behaviour of both soils' surface charge (andisol and ultisol) when MSM was sorbed. At pH lower than ZPC, hydrogen bonding has been suggested as interaction mechanism through the protonation of pyrimidine nitrogen moiety of SH and subsequent with the surface hydroxyls on the amorphous Fe oxide (positively charged), and also, these important adsorbents on VADS can retain SH through electrostatic attraction and ligand exchange.

In general, as the pH comes closer to the pK_a of herbicides, the sorption increases due to the hydrophobicity of their neutral form. It was also suggested that the hydrophobic part of the organic anion might sorb to a hydrophobic surface, with its polar end oriented towards the more polar aqueous phase. In VADS, a decrease in soil pH increases the net positive charge on the surface [30], at low pH the negatively charged chlorophenols are adsorbed onto the positive sites through electrostatic attraction. The Ca-bridging can take place between anionic SOM functional groups (e.g. carbolic and phenolic groups) or negatively charged sites on constant/variable-charge mineral surfaces and anionic pesticides (**Table 1**) [17, 31].

A large part, but not all, of the variation in herbicide sorption coefficients between different soils can be eliminated by expressing sorption on OC basis rather than on a total soil mass basis [24]. In this sense, K_d is usually normalised to the OC content of the soil ($K_{OC} = K_d/OC$) or to the OM content of the soil ($K_{OM} = K_d/OM$) [25, 31]. The K_d for a particular organic compound changes significantly from one soil to another, and generally increases as the OC content of the soil and the hydrophobicity of the chemical increases. In this sense, the OM has a high affinity for many non-polar pesticides and dominates their sorption in soils with more than 3% OM (**Table 1**). It reflects the fact that OC is the major sorption domain for non-ionisable herbicides on soils [32], whereas poorly crystalline minerals attract polar organic molecules.

6. Physical/chemical properties in QSAR models: a mechanistic interpretation

QSAR modelling is a useful technique to predict the activity of chemicals, such as pesticides in a short time and with low cost, establishing a statistical relationship between the activity of chemicals and their structural and physicochemical properties [34–37]. The development of QSAR models with regulatory purposes is based on precaution, while their application facilitates prevention. In this regard, the member countries of the organisation for economic co-operation and development (OECD) have developed a set of five guiding principles, enabling the practical application of QSAR modelling as a reliable tool in the regulatory context, which have been adopted by the European Union and United States [38–40]: (i) a defined endpoint; (ii) an unambiguous algorithm; (iii) a defined domain of applicability; (iv) an appropriate measure of goodness-of-fit, robustness and predictivity and (v) a mechanistic interpretation. The QSAR models are based on the assumption that chemicals are able to reach and interact with the target site by similar mechanism, related to their similar physicochemical properties [40].

6.1 QSAR models for sorption of pesticides in soils

The recent developments in QSAR models for the estimation of soil sorption coefficients of organic pesticides will be reviewed, taking into account the literature related to QSAR and its mechanistic interpretation. The aim of this section is to examine the descriptors used in the prediction of K_{oc} values on soil by means of QSAR models, which according to principle (v) should be associated with a 'mechanistic interpretation'. In this regard, two approximations have been used for the mechanistic basis of the models: (i) statistical approach and (ii) mechanistic approach. In the 'mechanistic approach', the descriptors selection can be guided by the modeller's a priori knowledge of the physicochemical properties involved in the mechanism proposed for the studied activity. Thus, presumed mechanistic meaning can be assigned to some molecular descriptor used; then, the modeller selects it personally from a limited pool of potential modelling variables, which are largely employed (for instance: $\log K_{ow}$, $\log P$, $\log D$, pK_a, OC, etc.). However, as the sorption mechanisms of pesticides on soils and especially on VADS are quite complex, their understanding is only possible at certain levels of approximation. The a priori selection of one (or more) physicochemical variable(s) for their mechanistic meaning, in relation to one assumed mechanism, is very risky. In this sense, important variables influencing other mechanisms, participating in such a response, could remain ignored.

6.2 Statistical approach: a posteriori mechanistic interpretation

The European joint research centre (JRC) QSAR models database provides information on the recent developments in QSAR models that can be used for purposes of regulatory assessment of chemicals (e.g. REACH registration). In this database, four detailed QSAR models to predict the sorption partition coefficient are normalised to the OC content of the soil (K_{oc}, $K_{oc} = K_d/OC$ or $K_{oc} = K_f/OC$). The QSAR Model Reporting Format (QMRF) documentation for all QSAR models is available in the JRC's QMRF Inventory [41–44]. All these models were developed by statistical approach, wherein no mechanistic basis for their descriptors selection was set a priori. The first was published few years ago [44]. In this study, QSAR model was developed using K_{oc} values of 142 non-ionic organic pesticides (34 split for the validation set). The K_{oc} data for 20 other chemicals were used as external validation set. The QSAR model developed in this study is represented by Eq. (1):

$$\begin{aligned} Log\, K_{oc} \;=\; & 0.96 - 0.26\; Polarity\ parameter\ (AM1)/distance \\ & + 1.07^{-0.02} ALFA\ polarizability\ (DIP)(AM_1) \\ & - 1.99\ Max\ net\ atomic\ charge\ (AM_1)\ for\ C\ atoms \\ & + 1.30^{-0.02}\ WNSA1\ Weighted\ PNSA\ \left(\frac{PNSA1 * TMSA}{1000}\right)(Zefirov) \end{aligned} \qquad (1)$$

N = 142 (nTr = 108, nPred =34); $R^2 = 0.75$; $Q^2_{LMO30\%} = 0.73$; $s^2 = 0.445$.

where '*ALFA polarizability (DIP) (AM1)*' and '*WNSA1 Weighted PNSA* $\left(\frac{PNSA1 * TMSA}{1000}\right)$ *(Zefirov)*' are descriptors that quantify the molecular size related to charge distribution. The '*Max net atomic charge (AM1) for C atoms* and the '*Polarity parameter (AM1)/distance*' are descriptors related to charges and to charge distribution.

Recently, Mansouri and Williams [41] published a new reliable QSAR model for estimating the K_{oc} of heterogeneous organic chemicals. This is freely available as an open-source, command-line application called OPERA (OPEn structure–activity/property Relationship App) [45]. The model was generated using 729 curated outlier-free experimental K_{oc} data, which were divided into training (545 compounds) and validation sets (184 compounds). The descriptors' selection consisted of coupling genetic algorithms (GAs) with the weighted *kNN* algorithm, which allows building a model with 12 molecular descriptors, related to *logP* and water solubility. These descriptors were calculated using the free and open-source software PaDEL [46]. The model has potential to correctly predict the $\text{Log}K_d$ values of organic pesticides since the predictivity-statistics obtained by external validation was quite significant, i.e. $R^2 = 0.71$ and RMSD = 0.61.

In both QSAR models exhibited above, compounds with large molecular size tend to have higher soil sorption than compounds with small molecular size, due to their lower water solubility. For electronic descriptors related to charges and charge distribution, the presence of active functional group adjacent to carbon allows a high charge on this atom, which together with likewise higher polarity leads to better water solubility. In the first, the minus sign in the QSAR equation (Eq. (1)) on these descriptors indicates that the higher the values, the lower the soil sorption. In the last update published by Gramatica et al. in JRC QSAR Model Database, QSAR Eq. (2) [43, 47] and Eq. (3) [42, 47], QSAR models were generated using the K_{oc} experimental data of 643 heterogeneous organic compounds obtained from literature. The median of the K_{oc} values was used when more than one value was available for a single compound.

$$Log\,K_{oc} = 0.87 + 0.26\,VP-0 - 0.23\,nHBAcc + 0.08\,nAromBond - 0.19\,MAXDP \quad (2)$$

where N = 643 (nTr = 93, nPred =550); $R^2 = 0.79$; $Q^2_{LMO30\%} = 0.79$; CCC (concordance correlation coefficient) = 0.89; RMSE = 0.54

$$Log\,K_{oc} = -1.92 + 2.07\,VED_1 - 0.31\,nHBAcc - 0.31\,MAXDP - 0.39\,CIC_0 \quad (3)$$

where N = 643 (nTr = 93, nPred =550); $R^2 = 0.79$; $Q^2_{BOOT} = 0.79$; CCC (concordance correlation coefficient) = 0.88; RMSE = 0.55.

The model described by Eq. (2) and Eq. (3) takes into account two different ways to describe a potential intermolecular adsorbate-adsorbent interaction. First, *nHBAcc* (Eq. (2) and Eq. (3)) is related to electronegative atoms of molecules that form hydrogen bond indicating a potential mechanism. Second, the *MAXDP* descriptor (Eq. (2) and Eq. (3)) related to molecule electrophilicity supposed a possible sorption mechanism of charge transfer between adsorbate and adsorbent. These mechanisms have been suggested for pesticides where amine and/or heterocyclic N atoms (e.g. AT) act as electron donors to acceptor structural groups of humic acid [48]. Moreover, Briceno et al. [14] conclude that the mechanisms involved in the AT soils' sorption are both hydrogen bonding and charge transfer (**Table 1**). The other two descriptors in each QSAR equation, Eq. (2) (*VP - 0* and *nAromBond*) and Eq. (3) (*VED1* and CIC_0), are related to molecular size and have positive signs. Models indicate in general that the larger compounds are more sorbed than leached. Consequently, the ability of these models in estimating the K_{oc} values of pesticides is restricted to sorption of non-ionised chemicals on permanent charge soils, which were widely represented in the calibration sets as well as in validation sets. Moreover, a mechanistic interpretation of molecular descriptors for non-ionisable pesticides was provided a posteriori, after modelling, by interpretation of the final model in view of an association between the descriptors used and the soil sorption predicted.

6.3 Mechanistic approach: polyparameter linear free energy relationships

In the last decade, the concept of polyparameter linear free energy relationships (PP-LFERs) has been widely used for the prediction of sorption coefficients of neutral organic chemical, because of its important mechanistic basis and good prediction power [49–53]. PP-LFERs are multiple linear regression (MLR) models that employ several solute- or sorbate-specific descriptors as independent variables and their fitting coefficients are denoted as they describe system-specific, solute-independent properties. In this sense, descriptors and their coefficients quantitatively describe the energetic contribution of different types of sorption coefficients [53, 54]. The major advantages of the PP-LFER approach are its solid mechanistic grounds and the use of uniformly measured calibration data.

In the last decade, different PP-LFER models for organic contaminants sorption on soil estimation have been proposed. Endo et al. [52] proposed two PP-LFER models at environmentally relevant concentrations. However, these models lack reliable PP-LFER descriptors for environmentally relevant chemicals (e.g. pesticides, pharmaceuticals and highly polar compounds, acids, bases, and ionic compounds). This deficiency also has been identified for PP-LFER models developed for high sorbate concentrations previously reported [55]. The PP-LFER models' reviews up to now mainly have been calibrated estimating $\log K_{oc}$ data of classical pollutants such as PCBs and PAHs and also of organic compounds that have chemical structure comparatively simple than chemicals of current environmental concern. These are often multifunctional or complex organic chemicals like pesticides and pharmaceuticals. The first reliable PP-LFER model for soil-water partitioning was calibrated with data from 79 polar and non-polar compounds that cover a more diverse and wider range of chemical classes than other PP-LFERs published. The model of Bronner and Goss [49] was validated using the experimental data for about 50 pesticides and pharmaceuticals not involved in the calibration set. This has potential to correctly estimate the K_{oc} data for multifunctional or complex organic chemicals like pesticides and pharmaceuticals. However, Sabljic and Nakagawa [53] suggest still important drawbacks to the general applicability of the developed model. In view of the scope of this section, we recommend the review made by Sabljic and Nakagawa [53] around this topic.

On the other hand, little attention has been paid to the general applicability of the calibrated PP-LFERs for predicting sorption to soils considering the diversity of soil mineralogy, variable surface charge, OC structures and their interactions [51]. The evaluation of possible applications of PP-LFERs in the study of partitioning of ionic organic chemicals is a subject of ongoing research [56, 57].

6.4 Mechanistic approach: QSAR models for sorption of ionisable pesticides

In the last decade, different authors developed equations to predict the sorption of ionisable and non-ionisable compounds in soils or sediments [25, 58–61]. Several models have expanded their applicability domain including soil properties and ionisation effects [48, 58, 59]. Franco et al. carried out a surface acidity correction, because the two units proposed by Bintein and Devillers [59] are dependent on soil properties, related to the surface potential of the colloid [25]. These researches suggest a general non-linear equation based on $LogK_{ow}$ for neutral and ionic species (a fragmentation of $LogD$) and the speciation of monovalent acids, monovalent bases and amphoteric species. Franco et al. aimed to predict pH-dependent K_d values of organic acids, considering speciation as a function of soil pH and species-specific partition equilibrium [60]. This modification of their previous models by replacing their constant terms pH_{opt} by a varying pH range allowed that the modified model performs significantly better than the original model for organic acids [25]. The two molecular descriptors, pK_a and $logPn$, and the two soil descriptors, OC and pH, used in the model have a major impact on the sorption of ionisable chemicals. Nevertheless, it was not successful to develop the analogous modified model for bases due to the contradictory effect of pH on the total sorption.

6.5 Physical/chemical properties of VADS: considerations for a mechanistic approach

In general, VADS are rich in OM, possess high specific surface area, variable surface charge and consequently pH-dependent CEC and AEC (anion-exchange capacity (AEC). In this way, they have significant differences with regard to soils with mineralogy dominated by constant charge minerals [2]. These differences in the physicochemical properties make difficult the K_{oc} prediction for organic ionisable compounds using generic QSAR models or also published PP-LFER models. In this regard, the descriptors selection to a mechanistic interpretation of pesticide-VADS system must be related to the type of organic compound (e.g. non-ionisable, ionisable, acid, base, etc.) and to physicochemical properties of VADS.

According to physical/chemistry characteristics of VADS, in these soils, the pH is a critical parameter when ionisable pesticide-VADS interaction is considered, because the sign and magnitude of total VADS surface charge depends on $pH_{equilibrium}$. The total VADS surface charge is defined by the ZPC_{Soil}. The total VADS surface charge, at a given pH, could be negatively charged if $pH_{equilibrium} > pH_{ZPC}$ affecting the pesticide-VADS interaction (i.e. andisol and ultisols), which is particularly important for ionisable pesticides. The Fe/Al oxide content is an important variable for ionisable pesticide sorption on positively charged active sites at $pH_{equilibrium} < pH_{ZPC}$ and also in the remaining oxide sites that exhibit positive charge even at pH higher than ZPC [4, 17, 48]. Additionally, it is necessary to consider the relationship between these oxides and OM due to blockage of specific sorption sites, and between oxides and pH due to the presence of pH-dependent sorption sites and the speciation of ionisable pesticides [48, 62]. In this regard, Hyun et al. [63, 64] demonstrated that anion exchange in VCS is significant for pentachlorophenol and prosulfuron. Moreover, the extent of anion exchange correlated well to the ratio of pH-dependent AEC to CEC (i.e. AEC/CEC) as well as the ratio of AEC to the total number of soil surface charge (AEC + CEC) (i.e. AEC/(AEC + CEC)). Caceres-Jensen et al. [5] studied the effect of MSM sorption on total VADS surfaces observing a change in total VADS surfaces produced when the highest MSM concentration was sorbed. A displacement of IEP_{Soil} to a higher pH was established for the soils. These results confirmed the contribution of charged surface sites on VADS to the sorption of anionic MSM through electrostatic interactions. Finally, soil composition, mineralogy (e.g. amorphous (hydro)oxides, Fe/Al oxide content), texture (e.g. silt, sand or clay content), surface area of colloids, OM, AEC and its relation with CEC (i.e. AEC/CEC, AEC + CEC, AEC/(AEC + CEC), ZPC_{Soil} and $pH_{equilibrium}$ are potential modelling VADS descriptors; due to these descriptors, the K_{oc} magnitude is strongly influenced by physicochemical properties of VADS, sorption sites and specific surface area [2, 17, 19, 63].

According to the type of organic compound, special attention has been given to ionisable organic compounds. With changes in the pH, the speciation of soil active sites and of the ionisable pesticide also change, affecting the sorption. For ionisable pesticides, $logD$ could be a good descriptor for the variation in hydrophobic interactions. Also, pesticide pK_a values are possible descriptors that take into account dissociation in order to describe the interactions for pesticides on VADS considering the soil pH. This molecular property is determinant of hydrophilic interactions for polar compounds [17, 48].

7. Conclusions

The surface charge amphoteric characteristics will confer to VADS physical/chemical properties absolutely different to constant-charge soils, where soil composition (i.e. SOM), mineralogy and variable charge are key components of most VADS, controlling soil sorption of INIH, representing an environmental substrate that may become polluted over time due to intensive agronomic uses. The *pseudo-second-order* model and TSNE have been the models that best describe the kinetics parameter and solute sorption mechanism, respectively, of INIH on VADS. These models are also necessary in order to develop and validate QSAR models to predict pesticide sorption on VADS to prevent potential contamination of water resources and predict environmental risks. In this regard, the last section of this chapter illustrates briefly some of the advances of QSAR models established for predicting the soils' sorption of pesticides with a focus on the mechanistic interpretation. In the generation of QSAR models, the statistical approach is the most used with a posteriori mechanistic interpretation, possibly due to complex sorption mechanisms of pesticides on soils. In the mechanistic approach (a priori mechanistic interpretation), few studies have paid attention to the diversity of soil mineralogy, texture, variable surface charge, OC structures and their implication on sorption of ionisable pesticides. Finally, the use of solute sorption mechanism models and QSAR models for pesticide sorption in soils will contribute to a better understanding of behaviour of pesticides on VADS.

Acknowledgements

This work was funded via projects DIUMCE FGI 36-17 from Universidad Metropolitana de Ciencias de la Educación, Chile, FONDECYT 11110421 from CONICYT, Chile and FB0807 (Basal Funding for Scientific and Technological Centers) from CONICYT, Chile.

Conflict of interest

The authors certify that they have no conflict of interest with the subject matter discussed in this chapter.

Author details

Lizethly Caceres Jensen[1*], Jorge Rodriguez Becerra[1] and Mauricio Escudey[2,3]

1 Laboratory of Physical and Analytical Chemistry, Department of Chemistry and Doctorate in Education Program, Universidad Metropolitana de Ciencias de la Educación, Santiago, Chile

2 Facultad de Química y Biología, Universidad de Santiago de Chile, Santiago, Chile

3 Center for the Development of Nanoscience and Nanotechnology, CEDENNA, Santiago, Chile

*Address all correspondence to: lyzethly.caceres@umce.cl

References

[1] Sarmah AK, Muller K, Ahmad R. Fate and behaviour of pesticides in the agroecosystem - a review with a New Zealand perspective. Australian Journal of Soil Research. 2004;**42**:125-154. DOI: 10.1071/sr03100

[2] Mauricio Escudey GG, Forster JE, Briceño M, Diaz P, Chang A. Chemical forms of phosphorus of volcanic ash-derived soils in Chile. Communications in Soil Science and Plant Analysis. 2001;**32**:601-616. DOI: 10.1081/CSS-100103895

[3] Báez ME, Espinoza J, Silva R, Fuentes E. Sorption-desorption behavior of pesticides and their degradation products in volcanic and nonvolcanic soils: Interpretation of interactions through two-way principal component analysis. Environmental Science and Pollution Research. 2015;**22**:8576-8585. DOI: 10.1007/s11356-014-4036-8

[4] Cáceres-Jensen L, Gan J, Báez M, Fuentes R, Escudey M. Adsorption of glyphosate on variable-charge, volcanic ash-derived soils. Journal of Environmental Quality. 2009;**38**:1449-1457. DOI: 10.2134/jeq2008.0146

[5] Caceres L, Fuentes R, Escudey M, Fuentes E, Baez MAE. Metsulfuron-methyl sorption/desorption behavior on volcanic ash-derived soils. Effect of phosphate and pH. Journal of Agricultural and Food Chemistry. 2010;**58**:6864-6869. DOI: 10.1021/jf904191z

[6] Brusseau ML, Rao PSC. The influence of sorbate-organic matter interactions on sorption nonequilibrium. Chemosphere. 1989;**18**:1691-1706. DOI: 10.1016/0045-6535(89)90453-0

[7] Cáceres-Jensen L, Rodríguez-Becerra J, Parra-Rivero J, Escudey M, Barrientos L, Castro-Castillo V. Sorption kinetics of diuron on volcanic ash derived soils. Journal of Hazardous Materials. 2013;**261**:602-613. DOI: 10.1016/j.jhazmat.2013.07.073

[8] Villaverde J, van Beinum W, Beulke S, Brown CD. The kinetics of sorption by retarded diffusion into soil aggregate pores. Environmental Science & Technology. 2009;**43**:8227-8232. DOI: 10.1021/es9015052

[9] Cáceres-Jensen L, Escudey M, Fuentes E, Báez ME. Modeling the sorption kinetic of metsulfuron-methyl on Andisols and Ultisols volcanic ash-derived soils: Kinetics parameters and solute transport mechanisms. Journal of Hazardous Materials. 2010;**179**:795-803. DOI: 10.1016/j.jhazmat.2010.03.074

[10] Qafoku NP, Ranst EV, Noble A, Baert G, Donald LS. Variable charge soils: Their mineralogy, chemistry and management. In: Adv. Agron. Elsevier: Academic Press; 2004. pp. 159-215. ISSN: 0065-2113

[11] Dahlgren RA, Saigusa M, Ugolini FC, Donald LS. The nature, properties and management f volcanic soils. In: Adv. Agron. Elsevier: Academic Press; 2004. pp. 113-182. ISSN: 0065-2113

[12] Shoji S, Takahashi T. Environmental and agricultural significance of volcanic ash soils. Global Environmental Research-English Edition. 2002;**6**:113-135

[13] Takahashi T, Shoji S. Distribution and classification of volcanic ash soils. Global Environmental Research-English Edition. 2002;**6**:83-98

[14] Briceno G, Demanet R, de la Luz Mora M, Palma G. Effect of liquid cow manure on andisol properties and atrazine adsorption. Journal of Environmental Quality. 2008;**37**:1519-1526. DOI: 10.2134/jeq2007.0323

[15] Báez ME, Fuentes E, Espinoza J. Characterization of the atrazine sorption process on andisol and ultisol volcanic ash-derived soils: Kinetic parameters and the contribution of humic fractions. Journal of Agricultural and Food Chemistry. 2013;**61**:6150-6160. DOI: 10.1021/jf400950d

[16] Espinoza J, Fuentes E, Báez ME. Sorption behavior of bensulfuron-methyl on Andisols and Ultisols volcanic ash-derived soils: Contribution of humic fractions and mineral-organic complexes. Environmental Pollution. 2009;**157**:3387-3395. DOI: 10.1016/j.envpol.2009.06.028

[17] Hyun S, Lee LS. Quantifying the contribution of different sorption mechanisms for 2,4-dichlorophenoxyacetic acid sorption by several variable-charge soils. Environmental Science & Technology. 2005;**39**:2522-2528. DOI: 10.1021/es048820p

[18] Vasudevan D, Cooper EM. 2,4-D sorption in iron oxide-rich soils: Role of soil phosphate and exchangeable Al. Environmental Science & Technology. 2004;**38**:163-170. DOI: 10.1021/es034135r

[19] Cea M, Seaman JC, Jara AA, Fuentes B, Mora ML, Diez MC. Adsorption behavior of 2,4-dichlorophenol and pentachlorophenol in an allophanic soil. Chemosphere. 2007;**67**:1354-1360. DOI: 10.1016/j.chemosphere.2006.10.080

[20] Grube A, Donaldson D, Kiely T, Wu L. Pesticides Industry Sales and Usage 2006 and 2007 Market Estimates. Biological and Economic Analysis Division Office of Pesticide Programs, Office of Chemical Safety and Pollution Prevention, U.S. Environmental Protection Agency Washington, DC; 2011

[21] Flores C, Morgante V, González M, Navia R, Seeger M. Adsorption studies of the herbicide simazine in agricultural soils of the Aconcagua valley, Central Chile. Chemosphere. 2008;**74**:1544-1549. DOI: 10.1016/j.chemosphere.2008.10.060

[22] Brusseau ML, Famisan GB, Artiola JF, Janick FA, Ian LP, Mark LB. Chemical contaminants. In: Environmental Monitoring and Characterization. Burlington: Academic Press; 2004. pp. 299-312

[23] Sabljic A. QSAR models for estimating properties of persistent organic pollutants required in evaluation of their environmental fate and risk. Chemosphere. 2001;**43**:363-375. DOI: 10.1016/S0045-6535(00)00084-9

[24] Katritzky AR, Kuanar M, Slavov S, Hall CD, Karelson M, Kahn I, et al. Quantitative correlation of physical and chemical properties with chemical structure: Utility for prediction. Chemical Reviews (Washington, DC, United States). 2010. DOI: 10.1021/cr900238d

[25] Franco A, Trapp S. Estimation of the soil–water partition coefficient normalized to organic carbon for ionizable organic chemicals. Environmental Toxicology and Chemistry. 2008;**27**:1995-2004. DOI: 10.1897/07-583.1

[26] Sparks DL. 5 - sorption phenomena on soils. In: Environmental Soil Chemistry. 2nd ed. Burlington: Academic Press; 2003. pp. 133-186. DOI: 10.1016/B978-012656446-4/50005-0

[27] Tan KL, Hameed BH. Insight into the adsorption kinetics models for the removal of contaminants from aqueous solutions. Journal of the Taiwan Institute of Chemical Engineers. 2017;**74**:25-48. DOI: 10.1016/j.jtice.2017.01.024

[28] Pizarro C, Fabris J, Stucki J, Garg V, Morales C, Aravena S, et al. Distribution

of Fe-bearing compounds in an Ultisol as determined with selective chemical dissolution and Mössbauer spectroscopy. Hyperfine Interactions. 2007;**175**:95-101. DOI: 10.1007/s10751-008-9594-z

[29] Stolpe NB, Kuzila MS. Relative mobility of atrazine, 2,4-D and dicamba in volcanic soils of south-Central Chile 1. Soil Science. 2002;**167**:338-345

[30] Sposito G. The operational definition of the zero point of charge in soils1. Soil Science Society of America Journal. 1981;**45**:292-297. DOI: 10.2136/sssaj1981.03615995004500020013x

[31] Hiradate S, Furubayashi A, Uchida N, Fujii Y. Adsorption of 2,4-dichlorophenoxyacetic acid by an andosol. Journal of Environmental Quality. 2007;**36**:101-109. DOI: 10.2134/jeq2005.0415

[32] Schuurmann G, Ebert R-U, Kuhne R. Prediction of the sorption of organic compounds into soil organic matter from molecular structure. Environmental Science & Technology. 2006;**40**:7005-7011. DOI: 10.1021/es060152f

[33] Báez ME, Rodríguez M, Vilches P, Romero E. Adsorption of methabenzthiazuron on six allophanic and nonallophanic soils: Effect of organic matter amendment. Journal of Environmental Science and Health, Part B: Pesticides, Food Contaminants, and Agricultural Wastes. 2001;**36**:95-105. DOI: 10.1081/PFC-100000920

[34] Reddy KN, Locke MA. Prediction of soil sorption (Koc) of herbicides using semiempirical molecular-properties. Weed Science. 1994b;**42**:453-461

[35] Sabljic A. On the prediction of soil sorption coefficients of organic pollutants from molecular structure: Application of molecular topology model. Environmental Science & Technology. 1987;**21**:358-366. DOI: 10.1021/es00158a004

[36] Wang Y, Chen J, Yang X, Lyakurwa F, Li X, Qiao X. In silico model for predicting soil organic carbon normalized sorption coefficient (Koc) of organic chemicals. Chemosphere. 2015;**119**:438-444. DOI: 10.1016/j.chemosphere.2014.07.007

[37] dos Reis RR, Sampaio SC, de Melo EB. An alternative approach for the use of water solubility of nonionic pesticides in the modeling of the soil sorption coefficients. Water Research. 2014;**53**:191-199. DOI: 10.1016/j.watres.2014.01.023

[38] Cronin MTD, Jaworska JS, Walker JD, Comber MHI, Watts CD, Worth AP. Use of QSARs in international decision-making frameworks to predict ecological effects and environmental fate of chemical substances. Environmental Health Perspectives. 2003;**111**:1391-1401

[39] Tropsha A. Best practices for QSAR model development, validation, and exploitation. Molecular Informatics. 2010;**29**:476-488. DOI: 10.1002/minf.201000061

[40] OECD. Guidance Document on the Validation of (Quantitative) Structure-Activity Relationship [(Q)SAR] Models. Paris: OECD Publishing; 2014. https://doi.org/10.1787/9789264085442-en

[41] Mansouri K, Williams A. QMRF-JRC-QSARDB: KOC model for the soil adsorption coefficient prediction from OPERA models. RN: Q17-26-0017; 2017. https://qsardb.jrc.ec.europa.eu/qmrf/endpoint. DOI: 10.13140/RG.2.2.27831.32163/1

[42] Gramatica P, Cassani S, Papa E. QMRF-JRC-QSARDB: QSARINS model 2 for log Koc. RN: Q17-26-0057; 2017. https://qsardb.jrc.ec.europa.eu/qmrf/endpoint

[43] Gramatica P, Cassani S. QMRF-JRC-QSARDB: QSARINS model 1 for log Koc. RN: Q17-26-0032; 2017. https://qsardb.jrc.ec.europa.eu/qmrf/endpoint

[44] M.m.d. team, QMRF-JRC-QSARDB: QSAR for the soil adsorption coefficient Koc. RN: Q13-26-0009; 2013. https://qsardb.jrc.ec.europa.eu/qmrf/endpoint

[45] Mansouri K, Grulke CM, Judson RS, Williams AJ. OPERA models for predicting physicochemical properties and environmental fate endpoints. Journal of Cheminformatics. 2018;**10**:10. DOI: 10.1186/s13321-018-0263-1

[46] Wei YC. PaDEL-descriptor: An open source software to calculate molecular descriptors and fingerprints. Journal of Computational Chemistry. 2011:**32**;1466-1474. DOI: 10.1002/jcc.21707

[47] Gramatica P, Cassani S, Chirico N. QSARINS-Chem: Insubria datasets and new QSAR/QSPR models for environmental pollutants in QSARINS. Journal of Computational Chemistry. 2014;**35**:1036-1044. DOI: 10.1002/jcc.23576

[48] Kah M, Brown CD. Adsorption of ionisable pesticides in soils. In: Reviews of Environmental Contamination and Toxicology. Vol. 188. New York: Springer; 2006. pp. 149-217

[49] Bronner G, Goss KU. Predicting sorption of pesticides and other multifunctional organic chemicals to soil organic carbon. Environmental Science & Technology. 2011;**45**:1313-1319. DOI: 10.1021/es102553y

[50] Card ML, Chin YP, Lee LS, Khan B. Prediction and experimental evaluation of soil sorption by natural hormones and hormone mimics. Journal of Agricultural and Food Chemistry. 2012;**60**:1480-1487. DOI: 10.1021/jf203093d

[51] Faria IR, Young TM. Comparing linear free energy relationships for organic chemicals in soils: Effects of soil and solute properties. Environmental Science & Technology. 2010;**44**:6971-6977. DOI: 10.1021/es100440a

[52] Endo S, Grathwohl P, Haderlein SB, Schmidt TC. LFERs for soil organic carbon-water distribution coefficients (KOC) at environmentally relevant sorbate concentrations. Environmental Science & Technology. 2009;**43**:3094-3100. DOI: 10.1021/es803157e

[53] Sabljic A, Nakagawa Y. Sorption and quantitative structure-activity relationship (QSAR). In: Non-First Order Degradation and Time-Dependent Sorption of Organic Chemicals in Soil. Washington, DC, USA: American Chemical Society; 2014. pp. 85-118. DOI: 10.1021/bk-2014-1174.ch005

[54] Endo S, Goss KU. Applications of polyparameter linear free energy relationships in environmental chemistry. Environmental Science & Technology. 2014;**48**:12477-12491. DOI: 10.1021/es503369t

[55] Nguyen TH, Goss K-U, Ball WP. Polyparameter linear free energy relationships for estimating the equilibrium partition of organic compounds between water and the natural organic matter in soils and sediments. Environmental Science & Technology. 2005;**39**:913-924. DOI: 10.1021/es048839s

[56] Abraham MH, Acree WE. Equations for the transfer of neutral molecules and ionic species from water to organic phases. The Journal of Organic Chemistry. 2010;**75**:1006-1015. DOI: 10.1021/jo902388n

[57] Zhang H, Shields AJ, Jadbabaei N, Nelson M, Pan B, Suri RPS. Understanding and modeling removal of anionic organic

contaminants (AOCs) by anion exchange resins. Environmental Science & Technology. 2014;**48**:7494-7502. DOI: 10.1021/es500914q

[58] Kah M, Brown CD. Prediction of the adsorption of ionizable pesticides in soils. Journal of Agricultural and Food Chemistry. 2007;**55**:2312-2322. DOI: 10.1021/jf063048q

[59] Binetin S, Devillers J. QSAR for organic chemical sorption in soils and sediments. Chemosphere. 1994;**28**:1171-1188. DOI: 10.1016/0045-6535(94)90335-2

[60] Franco A, Fu W, Trapp S. Influence of soil pH on the sorption of ionizable chemicals: Modeling advances. Environmental Toxicology and Chemistry. 2009;**28**:458-464. DOI: 10.1897/08-178.1

[61] Franco A, Trapp S. A multimedia activity model for ionizable compounds: Validation study with 2,4-dichlorophenoxyacetic acid, aniline, and trimethoprim. Environmental Toxicology and Chemistry. 2010;**29**:789-799. DOI: 10.1002/etc.115

[62] Villaverde J, Kah M, Brown CD. Adsorption and degradation of four acidic herbicides in soils from southern Spain. Pest Management Science. 2008;**64**:703-710. DOI: 10.1002/ps.1545

[63] Hyun S, Lee LS, Rao PSC. Significance of anion exchange in pentachlorophenol sorption by variable-charge soils. Journal of Environmental Quality. 2003;**32**:966-976. DOI: 10.2134/jeq2003.9660

[64] Hyun S, Lee LS. Factors controlling sorption of prosulfuron by variable-charge soils and model sorbents. Journal of Environmental Quality. 2004;**33**:1354-1361. DOI: 10.2134/jeq2004.1354

Chapter 7

Sorption of Heavy Metals on Clay Minerals and Oxides: A Review

Ifeoma Mary Ugwu and Onyedikachi Anthony Igbokwe

Abstract

Sorption of heavy metals plays a vital role in controlling environmental pollution. Here, we reviewed the sorption of heavy metals such as Ni, Co, Cu, Zn, V, Pb, Hg, In, As, Cd, Cr, Ga, Cs, Mn, V, Eu, Mo, Th, TI and Cr on metal oxides and clay minerals. The mechanism of association between these ions and the host minerals, and the factors controlling their sorption are discussed in detail. Both chemical and empirical methods of describing sorption mechanism are discussed. The sorption processes depend on the pH, metal concentration, ionic strength, temperature, time, adsorbent dosage, type of ion, surface area, type of adsorbent modification and nature of adsorbent. The review confirmed that both metal oxides and clay have capability of sequestering heavy metals, however, combination of both metal oxides and modified clay have enhanced capability of removing heavy metals from aqueous solution. These inorganic adsorbents have the regeneration and recycling potentials and can be used to remediate and sequester economic metals for commercial purposes, however, this needs future investigation.

Keywords: sorption, adsorption, heavy metals, remediation, oxides, clay minerals

1. Introduction

Heavy metal pollution is an environmental problem that has harmful effect on both aquatic and terrestrial environment. These metals are released to the environment through activities such as mining, electroplating and manufacturing of paper and pesticides in form of mine tailings or effluents [1]. They have ability to complex with minerals to form inorganic ligands with variable solubility and acid–base potentials, thus, making their remediation from contaminated land/soil a major concern [2, 3].

Pollution arising from heavy metals poses serious health problems to both human beings and animals [1]. *For example,* low concentration of cadmium in human being can lead to 'Itai-Itai' disease, lungs cancer, and kidney and liver problems [4, 5]. Ingestion of water contaminated with high level of Cr can cause ulcer, skin inflammation and pulmonary congestion [6, 7]. Health problems associated with ingestion or inhaling of nickel dust are diarrhoea, renal oedema, nausea, chest pain, pulmonary fibrosis and gastrointestinal ache. Excess of lead in the environment can lead to mental retardation, kidney and nervous disease, cancer [8, 9]. The Cu^{2+} is the most toxic form of copper [10], and health problems

IntechOpen

associated with copper are anaemia, headache, kidney damage and death [11, 12]. Large amount of zinc in human body can result to anaemia, fever, vomiting and skin inflammation [13]. Mercury ingestion can lead to damaging of reproductive and respiratory system, kidney and brain [14]. Also, health problem associated with arsenic are gastrointestinal problems, cancer, diabetes, and liver tumours [15]. Large concentration of Mn causes retardation of growth, infertility, fever, eye blindness and muscles [15]. Consumption of water with high concentration of cobalt can lead to heart problems, asthma, damage of thyroid, pneumonia, hair and weight loss, nausea and vomiting, [16]. Depending on the route and length of exposure to vanadium, it can lead to immunological, genetoxic, developmental, reproductive and systemic problems and even death [17]. Exposure to indium via respiration can lead to lung disease, kidney and lungs damage [18, 19]. Long term exposure of gallium may damage the kidney, liver, lungs and bone marrow [20]. Health effects of molybdenum are headache, weight loss, joint or muscle ache, diarrhoea, kidney and liver problems [21]. Exposure to radioactive caesium can lead to cells damage, cancer, diarrhoea, vomiting, nausea and mental abnormality in babies [22]. Exposure to thorium workers can cause lung and liver disease, cancer, liver diseases and death at large exposure [23].

Clay minerals and metal oxides are formed from weathering of primary minerals [24]. These minerals have high surface area and can absorb heavy metals from aqueous and natural environments [25]. Clay minerals, bentonite, sepiolite and palygorskite have been used extensively to remove heavy metals from wastewater and agricultural soil, and the mechanisms of remediation are sorption, precipitation and liming [26]. Heavy metal polluted water is mostly remediated using ion-exchange, adsorption and mechanism [27]. Even though, clay minerals have adsorption affinity for heavy metals, e.g. Cu and Zn, their roles are not significant when compared to iron oxides (e.g. goethite), manganese oxides (e.g. birnessite) and organic materials on a unit mass basis [28]. Comparison of the adsorption potentials of deep sea minerals such as clay minerals, Mn-oxides, Fe-oxides, calcite and apatite reveal that Mn and Fe oxides and oxyhydroxides are crucial phases for scavenging heavy metals [29] Lead readily sorbs on all phases but has greatest affinity for carbonate fluorapatite, however, Caesium (Cs) has affinity for illite [29]. In sea water, hydrated cations such as Mn^{2+}, Co^{2+}, Ni^{2+}, Zn^{2+}, Cu^{2+}, Ba^{2+}, $MnCl^{+}$ and $PbCl^{+}$ have strong affinity for Mn-oxide while HVO_4^{2-}, MoO_4^{2-}, $PbCO_3$ and $HAsO_4^{2-}$ preferentially adsorb onto Fe oxyhydroxide [29].

Sorption techniques have the ability to remove these contaminants via adsorption onto sorbents. However, the type of sorbent to be used for remediation depends on the sorption capacity, sorption pH and distribution coefficient of the contaminant [30]. The review on the sorption of Ni(II) adsorbents indicates that bisorbents are the most effective for removal of the ion from aqueous solution [31]. In addition, Pb and Cd be successfully immobilised with wide range of low cost materials such as metal oxides, animal manure, apatite, lime, biosolids, and biochar [32]. However, the review on the adsorption capabilities of low-cost sorbents such as agricultural waste, household wastes, industrial by-products and sludge indicates that even though these material has great adsorption potentials their particle size, surface area, contact time, initial concentration of ion, adoption dose are not stated by the researchers [33]. This poses great limitation for the use of these adsorbents for remediation. Further review of the sorption of nickel, copper, lead, cadmium, caesium, chromium, europium and thorium on both natural and synthetic materials reveal that more studies on multi-component sorption of these metals are required [34]. Therefore, synthetizing effective and low cost adsorbent that can be recycled for removal of contaminants from aqueous environment will be of high interest

[34]. Similarly, synthesising and discovering new novel methods of using modified clay for environmental remediation is required metals [25].

To this end, this paper presents the review the sorption of heavy metals such as Ni, Co, Cu, Zn, V, Pb, Hg, In, As, Cd, Cr, Ga, Cs, Mn, Eu, Mo, Th, TI and Cr on both natural and synthetic metal oxides and clay minerals for soil and water remediation. The main aim of this review is to summarise the adsorption mechanism and recent remediation method of single and multicomponent system of these toxic metals using clay minerals and metal oxides. In addition, the advantages of reusability these adsorbents after remediation are discussed.

2. Adsorption mechanism and models

2.1 Adsorption mechanism

The interactions between metallic ions and mineral surfaces in any environment allow the sorption of ions to the solid surface, thereby decreasing the concentration of ions in the aqueous phases below the solubility limits of the solid phase [35]. Determination of partitioning of heavy metals among the solid phase and accurately predicting their mobility in natural environment requires adequate knowledge of the chemical process such as ion exchange, adsorption, surface precipitation, coprecipitation and mineralisation, that govern sorption mechanism [36]. For example, cobalt sorbs to goethite through surface complexation, surface precipitate and structural incorporation depending on the concentration and sorption duration [37].

Surfaces of substances (oxides, organic and inorganic) have surface functional groups or hydroxyl groups of solute ligands [38]. These functional groups are dominant in natural environments as oxides of Al, Fe and Si and in aqueous solution. The surface charge of metallic oxy-hydroxides, phosphate and carbonates is formed through ionisation of surface groups [39]. The pH of surface oxide and oxyhydroxides changes due to adsorption of protons or ions from solution [40]. Thus, the surface charge set up potential differences between the sorbent and sorbate, thereby influencing the approach of ions towards the surfaces [40]. Consequently, surface hydroxyl groups are responsible for complexation on metal oxides, oxyhydroxides, hydroxides and aluminosilicates. The type of surface functional group influences the sequence of the adsorption reaction, variation in adsorption solution chemistry, electrical properties of the interface and the adsorption ability [41]. The reaction of metals and oxygen atoms at the surface results in the formation of hydroxyl groups (OH^-), which are taken as part of the surface instead of solution. The surface hydroxyl is amphoteric (i.e. they can either accept a proton or donate a proton) as shown below:

$$\equiv S\text{–}OH + H^+ \rightleftharpoons \equiv S\text{–}OH_2^+ \quad (1)$$

$$\equiv S\text{–}OH \rightleftharpoons \equiv S\text{–}O^- + H^+ \quad (2)$$

Adsorption of metal to the surface takes place via substitution of surface protons (H^+) to form inner sphere complexes:

$$\equiv S\text{–}OH + M^{Z+} \rightarrow \equiv S\text{–}O\text{–}M^{Z-1} + H^+ \quad (3)$$

$$2{\equiv}S{-}OH + M^{Z+} \rightarrow ({\equiv}S{-}O)_2M^{Z-1} + 2H^+ \tag{4}$$

$$2{\equiv}S{-}OH + M^{Z+} \rightarrow ({\equiv}S{-}O)_2M^{Z-1} \tag{5}$$

The surface oxygen, ΞSO$^-$ may attract ions from bulk solution to form outer sphere complexes [42]:

$$\equiv S{-}O^- + M(H_2O)_6^{2+} \rightarrow \equiv S{-}O^-........(H_2O)_6M^{2+} \tag{6}$$

Bidentate ligands (Lewis bases, e.g. Cl^-, OH^-, H_2O, NH_3^+) are sorbed to the surface by the replacement of the surface hydroxyl group (ΞSOH) as follows:

$$\equiv S{-}OH + L^- \rightarrow \equiv S{-}L + H_2O \tag{7}$$

Adsorbed ligands can also take up another metal while metals already adsorbed can attach themselves to another ligand [43].

$$\equiv S{-}L + M^{Z+} \rightarrow \equiv S{-}L{-}M^{Z+1} \tag{8}$$

$$\equiv S{-}O{-}M^{Z-1} + L^- \rightarrow \equiv S{-}O{-}M{-}L^{Z-2} \tag{9}$$

2.2 Adsorption and Kinetic models

An adsorption isotherm is a relationship between the amount adsorbed and the concentration of the adsorbate at a constant temperature [39]. Both empirical and chemical models are used in describing the adsorption reaction. The former model uses mathematical expressions to interpret adsorption (e.g. Freundlich isotherm, Langmuir isotherm and distribution coefficient) whereas the latter uses a molecular approach to calculate equilibrium constants (e.g. constant capacitance model, triple layer models) [44]. Langmuir isotherm is based on the assumption that only one monolayer is formed during reaction, presence of equivalent sites, immobility of adsorbate and absence of adsorbate-adsorbent interaction [39]. The difference between Langmuir isotherm and Freundlich is that Langmuir isotherm considers maximum sorption whereas Freundlich considers infinity site [39]. However, adsorption isotherms are of limited application when modelling because of the complex and variability of natural environment with heavy metal and existence of sorbents with pH dependent surface charges [39]. Therefore, application of surface complexation models that can integrate both the experimental data and the electrostatic interaction between the ion and the oxide surface is required for modelling the mobility sorption of heavy metal. Computer models such as MINTEQ, FITEQL EQLFOR, HYDRAQL and SOILCHEM can calculate both equilibrium speciation and constants [44–46].

Adsorption kinetic model is used to investigate sorption mechanism as well as the rate controlling steps in a batch reaction [34]. Kinetic studies is carried out in a way to vary the initial concentration of adsorbate, temperature, time, sorbent dosage, particle size, pH, type of sorbent and sorbate [34]. It is a continuous measurement of experimental data until equilibrium is attained. The result is tested using kinetic equations to determine the best fit, which will give a better understanding of the sorption mechanism. The adsorption isotherms

Isotherm equations	Kinetic equations
$q_e = \frac{RT}{b} \ln (aC_e)$ Temkin	$\frac{dq}{dt} = k(q_e - q)^2$ Pseudo-second-order model
$\theta = \frac{K_{ad}[M]}{1+K_{ad}[M]}$ Langmuir	$\frac{dq}{dt} = k(q_e - q)$ Lagergren model
$\theta = K_{ad}[M]^n$ Freundlich	$\frac{dq}{dt} = \alpha \exp(-\beta q)$ Elovich model
$q_e = q_{\max} \exp\left(-\left(\frac{RT\ln\left(\frac{C_e}{C_s}\right)}{\beta E_0}\right)^2\right)$ Dubinin-Radushkevich	$\frac{dq}{dt} = k_1 C - K_2 C_s$ Bhattacharya and Venkobachar model
$q_e = q_{\max} \frac{BC_e}{q_{\max}(C_e - C_s)[(1+B-1)(C_e/C_s)]}$ Brunauer, Emmett and Teller (BET)	$\frac{dq}{dt} = k_{sorp} C(q_e - q) - k_{des} q$ Adam-Bohart-Thomas
$\log \frac{\theta}{C_0} = \log K_{FH} + n_{FH} \log(1-\theta)$ Halsey	
$q_e = \frac{K_{RP} C_e}{1 + a_{RP} C_e^{\beta}}$ Redlich-Paterson	
$q_e = q_{\max} \frac{(K_s C_e)^{\gamma}}{(1+(K_s C_e)^{\gamma})}$ Sip	
$q_e = q_{\max} \frac{b_T C_e}{\left(1+(b_T C_e)^{nT}\right)^{1/n_T}}$ Toth	

θ = fraction of surface coverage; [M] = concentration of ion; Ce = concentration at equilibrium; Cs = solubility of adsorbate at a particular temperature; q_{max} = saturated monolayer sorption capacity; qe = the amount of solute adsorbed at equilibrium; q = the amount of solute adsorbed at any given time 't' #; C = the concentrations of sorbate in solution at any given time 't'; Cs = the concentrations of sorbate in sorbent at any time 't'.

Table 1.
Adsorption isotherm and kinetic model equations.

and kinetics commonly used are listed in **Table 1**. In addition, most of the adsorption of ions on clays and oxides are best described by Langmuir, Dubinin-Radushkevich and Freundlich isotherm models and Pseudo-second-order kinetic model [47–53].

Better understanding of sorption mechanism is done using technologies and spectroscopy studies such as Attenuated Total Reflection Infrared [37, 54], Fourier Transform Infrared (FTIR) spectroscopy, extend X-ray absorption fine structure spectroscopy (EXAFS) [55–57], powder X-ray diffraction (XRD), X-ray photoelectron spectroscopy (XPS) and Electron paramagnetic resonance (EPR) [58]. Spectroscopic studies show that copper, nickel, zinc and cobalt bind to Mn(IV) vacancy sites in birnessite to form inner-sphere surface complexes via interlayer triple-corner-sharing or enter the vacancy site to become part of the structure [56]. ATR-FTIR is very useful in distinguishing the outer-sphere and inner-sphere complexes on minerals. Yoon et al., determined of the types and structures of the adsorption complexes formed by oxalate at boehmite (γ-AlOOH)/water and corundum (α-Al_2O_3)/water interfaces using in situ ATR-FTIR spectroscopy and quantum chemical [54]. Their findings suggest that the adsorption mechanism involves loss of protons from aqueous species during ligand-exchange reaction and is useful clue on the transport mechanism and removal of toxic elements in natural water.

3. Investigations of the sorption properties of clay and metal oxides

Clay minerals, e.g. montmorillonite, illite, and kaolinite, are known to have ability of adsorbing heavy metals in their inter layers through ion exchange reaction [25]. They generally have high sorption and expansion properties and are widely used for removal of contaminants from aqueous solution [25]. Their specific roles for adsorption of heavy metals from aqueous solutions and natural environments are discussed below. These adsorbents and the metals they are capable of removing from aqueous solution are summarised in **Table 2**.

Metals	Adsorbents	Temperature (°C)	Concentration	Optimum recovery pH	Time for maximum adsorption	References
Zn(II) Cu(II) Mn(II) Cd(II) Pb(II) Ni(II)	Montmorillonite	25–26.85	500 mg/L	6–8	180 min	[47, 62, 63]
U(VI) Th (IV)	Jordanian Bentonite	25	100mg/L	3	18 h	[49]
Cu(II) Ag(I)	Verde-lodo bentonite	Cu: 60 Ag: 20	1.2429 mmol/L	<4.5	2 days	[75]
Th (IV)	Activated bentonite					[76]
In(II)	Chitosan-coated bentonite	25	70 mg/L	4.0	300 min	[77]
Cs(II)	Mixture of iron pillared Layered montmorillonite (80%) and goethite (20%)	15	13.3 mg/L	7.3.	30 min	[50]
Cu(II) Cs(II)	Aluminium-pillared-layered montmorillonites	—	100 mg/L	For Cu 4.0–6.0 and For Cs 3.0–8.0	1.5 h	[51]
Cr (VI))	Magnetite nanoparticles	25	50 mg/L	3.0	3 h	[53]
Mn(II) Ni(II) Cu(II)	Hexagonal birnessite	25	143 mg/L	7–8	24 h– 22 days	[55] [100]
Pb(II) Cu(II) Zn(II) Cd(II)	Hexagonal birnessite Hydrous Mn-oxide (HMO)-coated clay	25	2801 mg/L	4.5 to ~5.5	24 h– 2 weeks	[101] [102]
Hg(II) Cr(III) Pb(II) Cu(II) Zn(II) Ba(II) Ni(II) Mn(II) Cd(II) Ba(II) Ag(I)	Ca-montmorillonite	—	1 N	4.8	16 h	[64]

Metals	Adsorbents	Temperature (°C)	Concentration	Optimum recovery pH	Time for maximum adsorption	References
Eu(III)	Na-montmorillonite	25	1×10^{-3} mol/L	~7	8 days	[65]
Cd(II)	Heated ball clay Combination of montmorillonite and humic acid	30	50 mg/L	8 8.5	18 h 24 h	[68] [111] [112]
Pb(II) Cd(II) Ni(II) Cu(II)	Kaolinite	30	160 mg/L	7	30 min	[69]
Cd(II) Co(II) Cu(II) Pb(II) Ni(II) Cr(VI)	Acid-activated form kaolinite	30	250 mg/L	Co and Ni: pH 7.8 Cu and Pb: Ph 5.8 Cd: Ph 10, Cr: pH 7	40–240 min	[70] [71]
Cs(II)	Clay soil with predominantly illite as clay mineral	25	4 mg/L	8	16 h	[72] [73]
TI(I)	Illite	25	8.39×10^{-6} mol/L	7.18	6 days	[74]
Pb(II) Cu(II) Ni(II)	Synthetic magnesium-aluminium layered double hydroxides modified palygorskite	18–40	500 mg/L	5.0	4.0, 8.0, and 20.0 min, for Pb(II), Cu(II) and Ni(II), respectively	[84]
Pb(II)	Palygorskite-iron oxide nanocomposite	25			12 h	[85]
As(III) Cd(II) Cr(III) Cu(II) Hg(II) Ni(II) Pb(II)	Fe-oxide modified smectites (montmorillonite and saponite)	25	100 μg/L	pH 5	2 h	[88]
Ga (III) V(V) Co(II) Ni(II)	Goethite	20–25	10–58 mg/L	pH 7	24 h	[38] [93] [94] [98]
Cu(II) Zn(II) Cd(II) Pb(II)	Amine-magnetic Fe-oxide	30	1.0 mmol/L	6	60 min	[103]
Ni(II)	Mixture of goethite, humic acid and kaolinite	30	100 mg/L	5.5	2 h	[104]

Table 2.
Summary of metals, potential adsorbents and experimental parameters.

3.1 Montmorillonite clay

The sorption ability of montmorillonite and its remediation potential for nickel and copper from polluted water was investigated by [47]. In their study, sorption rate follows pseudo-second-order rate model and the adsorption model fits to Langmuir model. Also, the point of zero charge (PZC) for the aqueous solution containing both metal at different ionic strengths 0.01 and 0.001 M KCl, is pH 3.4 $\pm$ 0.2 [47]. In contrast, the sorption of cadmium, chromium and lead (II) from wastewater on clay follow 1st order kinetics and can be modelled with both Freundlich and Langmuir model [59]. This study reveals that clay has the capability of removing these metals from aqueous solution at low temperature. In another study, the adsorption of Pb on clay from Tunisia reached maximum at pH 7 and declined above pH 7 due to formation of Pb precipitates, and the mechanism of adsorption via ion exchange [60]. Similarly, sorption of Cu(II) and Ni(II) onto white montmorillonite from obtained Kahramanmaras show increased adsorption with increasing amount of the initial amount of adsorbent, pH and temperature. The white montmorillonite shows more affinity for Cu(II) with 100% adsorption compared to Ni(II) [61]. In another study, sorption of Ni(II) on montmorillonite is influence by the presence of copper in a bi-metal aqueous solution with distortion of the shape of the adsorption isotherm whereas the presence of nickel has negligible effect on the sorption of copper [58]. EPR and XAS analyses of adsorption data reveal that both copper form inner-sphere complexes at pH 8 in both single-metal and binary metal system. However, nickel in single metal system forms inner-sphere complex at pH 8 but in binary-solute systems, it forms Ni phyllosilicate co-precipitate/alpha-$Ni(OH)_2(s)$ precipitate at pH 8.0 [58]. In another study, the sorption of Ni(II) and Mn(II) on montmorillonite from Ugwuoba, Nigeria, reveals 90% removal of these ions from solution containing 500 mg/L at pH 6.0 with maximum adsorption capacity of 166.67 mg/g for nickel and 142.86 mg/g for manganese [62]. Acid treated montmorillonite has enhance adsorption ability for heavy metals (Zn, Cu, Mn, Cd, Pb, Ni), depending on the initial metal concentration [63].

Similarly, the adsorption of mercury, chromium, lead, copper, zinc, barium, nickel, manganese, cadmium, silver on Ca-montmorillonite from single metal solutions followed second-order at greater rate compared to montmorillonite [64]. Copper, lead and chromium forms surface precipitation with montmorillonite whereas mercury, zinc, barium, nickel manganese, cadmium and silver sorb onto montmorillonite via adsorption and ion exchange to form both inner and outer sphere complex at the silanol and aluminol site [64]. However, sorption of bi-metal solution of these metals is better described with Langmuir isotherm model [64]. In another study, Europium (III) sorbs on Na-montmorillonite to form an outer-sphere complex at the exchange site as well as inner-sphere surface complexes at the 'aluminol' and 'silanol' edge sites [65]. In another study, heavy metals such as Hg(II), Cr(III), Pb(II), Cu(II), Zn(II), Ba(II), Ni(II), Mn(II), Cd(II), Ba(II) and Ag (I) in a single metal system adsorbs more on Ca-montmorillonite compared to Na-montmorillonite [66]. The adsorption selective sequence is Hg(II) > Zn(II) > Ba (II) > Cd(II) > Ni(II) > Mn(II). However, Na-Montmorillonite is more effective for removal of Pb(II), Cu(II), Co(II), Cd(II) and Zn(II) from aqueous solution than Ca-Montmorillonite [66]. In general, montmorillonite clay and its modified form has stronger adsorption capacity for arsenic, cadmium, chromium, cobalt, lead, iron, manganese, nickel and zinc than kaolinite and its modified form [67].

3.2 Kaolinite

Adsorption of Cd, Cu, Ni, Zn, Pb and Cr with Ball clay (kaolinite with high plasticity) at 30°C indicates that it sorbs more ion compared to Ca-montmorillonite, illite, kaolinite and Kaolin, and has the higher affinity for cd relative to other metals [68]. The sorption was at the maximum at pH 6 and 100% adsorption of 50 mg/L Cd. The sorption rate follows pseudo-first-order kinetics whereas the Langmuir model is used to describe the isotherm model [68]. Similarly, Pb(II), Cd(II), Ni(II) and Cu(II) sorption on kaolinitic clay from Longyan, China, show maximum sorption in 30 min and reduction of Pb concentration from 160.00 to 8.00 mg/L [69]. Another study found that adsorption of Cd(II), Co(II), Cu(II), Pb(II) and Ni(II) onto Acid treated kaolinite have great potential for removal these metals from aqueous solution compared to untreated kaolinite [70]. However, the use of natural and modified kaolinite clays for the removal of Cr(VI) for contaminated water is pH dependent and the adsorption increases from pH 1–2 at equilibration time of 240 min [71].

3.3 Illite

The adsorption of Caesium on clay made up of predominately of illite in the presence of small amount of organic matter and Fe-Oxide indicate that presence of organic matter play significant role for Cs uptake. The adsorption is described by Freundlich [48]. Maximum sorption of Caesium on to clay soil occurs at pH 8 and ambient temperature [72]. Comparative study of Cs sorption on illite, montmorillonite and kaolinite show that Cs is most reactive and has strong affinity for illite compared to other clay [73]. The sorption data was successfully modelled with 1-pK Diffuse layer model [73]. In another study, thallium is found to associate with illite in natural environment and sorption studies reveal that TI(I) sorbs more onto illite than smectite [74]. The affinity for TI follows this sequence: MnO_2 > illite > smectite = ferrihydrite > = Al_2O_3 = goethite > SiO_2. However, in presence of Rb, Cs, Ti adsorption to illite is less [74].

3.4 Bentonite

Removal of silver and copper in a binary solution containing the two ions with Verde-lodo bentonite reveals that the adsorbent has high adsorption affinity for copper than silver [75]. Cu adsorption on Verde-lodo bentonite is more at an elevated temperature and attained adsorption capacity of 0.110 mmol/g at 60°C, whereas silver attained 0.090 mmol/g at 20°C [75]. In another study, adsorption of indium on chitosan-coated bentonite is best described by Langmuir isotherm whereas the adsorption kinetics fits the pseudo second order [76]. Thorium (IV) sorption onto activated bentonite depends on temperature, pH, ionic strength, and type of anion and the adsorption kinetics can be described by pseudo-second-order model [77]. The activated bentonite is effective for removal of thorium (IV) and associates with it via surface complexation [77]. In the another study, bentonite was effectively used to remove thorium (IV) at pH 3, equilibration time of 18 h and 25°C and adsorption was fitted to Freundlich, Langmuir and Dubinin-Radushkevich isotherm models. Desorption experiment shows that thorium (IV) is best recovered with 1.0 M HNO_3 [49]. However, the use of bentonite for sorption of heavy metals has some negative effect. This is because increase in concentration of heavy metals such as zinc, lead and copper results to decrease in liquid limit, swelling potential, swelling pressure and free swelling of bentonite but increases its hydraulic conductivity due to sorption of heavy metal in the double layer structure [78].

3.5 Natural clay

Competitive adsorption and desorption cadmium, chromium, copper, lead and zinc on natural clay show difference in the optimum maximum pH of adsorption and metal affinity for single and multi-element adsorption [79]. The adsorption sequence for single metal sorption is Cr > Pb > Cu > Cd > Zn whereas Cr > Cu > Pb > Cd > Zn is the obtained sequence in the multicomponent system. The study also notes that the initial pH determines the solubility and removal of heavy metal in competitive multi element scenario [79]. In both the single and multielement adsorption, about 90–100% metal uptake expect Zn was achieved at pH 2–12 [79]. In another study, sorption of Cd, Cu, Pb, and Zn with concentration of 10 mmol/L on clay minerals, Fe rich clay minerals, clay-Fe oxide, Fe-oxyhydroxides, and calcite at both alkaline and acidic condition show that cadmium and zinc are adsorbed in acidic media by clay minerals whereas copper and lead sorb more at alkaline meda [80]. Sorption of Pb(II), Cd(II), Cu(II) and Zn(II) on clay obtained from Aleg formation, Tunisia that has surface area of 71 m^2/g indicate adsorption sequence of Pb(II) > Cu(II) > Zn(II) > Cd(II) and capacity of 131.58 mg/g in single metal systems and less than 50.10 mg/g for multi-metal [81]. Another study found that Cd(II) adsorption is effective for clay samples heated at 200°C, however, above 200°C adsorption decreases as a result of loss of SiOH or Al–OH binding sites [68]. Similarly, clay from china shows to be effective for removal of Cd at 30°C [82]. In addition, investigation for the temperature and pH effect on Cr (VI) adsorption reveal that sorption rate increases with increasing temperature and decreasing pH for concentration range of 0.743–1.422 mg/g [83].

3.6 Palygorskite

Synthetic magnesium-aluminium layered double hydroxides modified palygorskite shows high adsorption capacity for of Pb(II), Cu(II) and Ni(II) in the aqueous solution with initial concentration of 100 mg/L at pH 5.0 [84]. The reaction fit into Langmuir isotherm and the pseudo-second-order kinetic model [84]. In another study, the use of synthetic palygorskite-Fe oxide nanocomposite removal of Pb(II) from aqueous solution with initial 20–500 mg/L Pb indicates that the adsorbent has maximum Pb(II) adsorption capacity of 26.6 mg/g at 5 g/L adsorbent 150 min^{-1} stirring speed, and 25°C [85]. The adsorption data best fitted to the Langmuir isotherm model and pseudo-second-order kinetic model [85].

3.7 Modified clay

Montmorillonite clay modified with tetramethylammonium cations show enhances adsorption capacity for copper compared with unmodified montmorillonite, and removed up to 99.4% Cu(II) from aqueous solution with adsorption capacity of 925.93 mg/g [86]. Another study found that modification of clay minerals with hexadecylammonium bromide (HDTMA) is effective for attenuation of As(III) and As(V) in the presence of Cd(II), Cu(II) and Mn(II) and anions such as glycine, iminodiacetic acid, ethylenediaminetetraacetic acid disodium salt, oxalic acid, phosphate and sulfate [87]. In contrast, Fe-oxide modified smectites (montmorillonite and saponite) have the best adsorption capacity for removal of As (III), Cd(II), Cr(III), Cu(II), Hg(II), Ni(II), Pb(II) and Zn(II) from aqueous solution with at concentration of 10–100 ppb [88]. Mercury is sequestered in Al pillared Clay from Aleg formation, Tunisia at pH 3.2 and 240 min equilibration time [89]. The adsorption fit to Langmuir model with adsorption capacities 49.75 mg/g whereas the adsorption kinetics fits to pseudo-second-order kinetic. The adsorption reaction is exothermic [89]. Another study found that Indium sorbs readily on montmorillonite Metosol modified with di(2-ethylhexyl) phosphoric acid with kinetics that follow pseudo-first- and pseudo-second-order models [89]. Sorption of copper and caesium on Al-pillared montmorillonites follow a pseudo-first-order equation and pH of sorption 4.0–6.0 for copper and 3.0–8.0 for caesium [51].

Caesium sorption is best described with Langmuir model while copper fits the Freundlich isotherm model [51]. Sorption of Cs onto Fe-pillared montmorillonite, goethite and their mixture show that optimum sorption pH range is 5–9 and 93% removal of Cs occurred with the mixture of 20% goethite with Fe-pillared montmorillonite. The sorption process is exothermic and follows pseudo-first-order kinetics and can be described with Dubinin-Radushkevich, Freundlich and Langmuir models [50]. Also, combination of Al-pillared montmorillonite and Fe_3O_4 an MMT/Fe3O4 removes Cs from contaminated soil and water and can be recycled efficiently, and can be applied for the remediation of Cs contaminated land. The adsorption follow pseudo-second-order kinetics and Freundlich isotherm model [52]. In addition, maximum adsorption of Cr(IV) from aqueous solution was achieved by using clay heated at 200 to 400°C [91]. The heating process increased the exposure of the adsorptive site and the adsorptive capabilities [91].

3.8 Metal oxides

The Fe^{3+} oxides and Mn-oxides such as goethite, magnetite and birnessite are usually used in sorption studies of heavy metals and anions because of its simple synthesis in the laboratory [40, 92]. These metal oxides act as hosts for toxic trace elements with valences of +2, +3 and +5 [92]. The potential use of these minerals and their mixtures for remediation of contaminated land are reviewed below.

3.8.1 Goethite

Gallium adsorbs on to goethite to form Ga hexa-coordinated >FeOGa (OH)(2)(0) surface complexes at both acid and alkaline pH and exhibit the same isotope fractionation [93]. In another study, vanadium (V) adsorbs onto goethite to form bidentate corner-sharing complex from initial solution of 2.5 and 25 mg/L at pH 1.5 to 12 [94]. Mononuclear V(V) complexes are present at 25 ppm whereas both mononuclear and polynuclear V(V) complexes exist at 25 mg//solution species form [94]. In another study, at concentration < 10 mg/L and pH 7, cobalt sorbs on goethite at pH 7 to form bidentate edge and corner complexes but form surface precipitates, bidentate edge and corner complexes from 10 to 58 mg/L and pH 7.5 [37]. With ageing, Co sorption on goethite is irreversible depending on the initial concentration [37]. In addition, sorption of Pb, Hg, Cd, Zn, Cu, Ni, Co, Mn, Cr and Al on goethite from solution 10^{-6} M was achieved in 2 h and 20°C, but different pH [94]. The pH of 50% sorbed metals (pH 50) ranges from 2.81 for Hg to 6.45 for Mn [95]. The adsorption of Cu(II) with initial concentration of 25 mg/L onto goethite, haematite and lepidocrocite from pH 2–7 reveal that at pH of total metal sorption is 6 for goethite, 6.2 for haematite and 6.8 for lepidocrocite [56]. In same study, EXAFS spectra show that Cu(II) adsorbs these Fe-(hydr)oxides to form two or three bidentate edge sharing, corner sharing and tridentate complexes [56]. Adsorption of Ni, Zn, and Ca onto the goethite reveal that goethite has stronger affinity for Ni and Zn compared to Ca, however, in no observable competition occur in Ni–Ca and Zn–Ca bi-metal systems [96]. The adsorption is well described with single-site Langmuir model [96]. In another study, indium with concentration of 6–29 μg/L was immobilised form acid mine drainage by raising it pH to 8 and subsequent sorption onto iron oxide [97]. In addition, up to 100 mg/L Ni sorbs onto goethite at pH 8, 25°C and equilibration time of 21 days [98].

3.8.2 Birnessite

Mn(II) forms edge-sharing complex with birnessite whereas Ni(II) forms triple corner-sharing complexes at pH 6.5–7.5, however, the addition of Mn(II) in

Ni(II)-birnessite suspension at pH 6.5 results to formation of edge-sharing Ni(II) complexes due to site competition [99]. However, at pH 7.5, the presence of Mn(II) results to transformation of birnessite into feitknechtite that encourages sorption and incorporation of Ni(II) from solution [99]. This suggests that alteration of birnessite can influence the solubility of nickel in anaerobic environment [98]. In another study, hexagonal birnessite (δ-MnO_2) was used to sorb Cu(II) containing 143, 77 and 32 mg/L at pH 1–9 [100]. At pH 3, 100% adsorption results to about 5, 2.5 and 1% of copper in the adsorbate [100]. However, EXFAS characterisation of adsorbates at pH 4 reveals that Cu forms by inner-sphere complexation whereas at pH 8, it associates with birnessite via structurally incorporation [100]. Sorption of Zn onto synthetic δ-MnO_2 from an initial solution containing 2000 mg/L of Zn occurs at pH ~ 1 and at maximum pH ~5 [101]. However, desorption reaction indicates the Zn sorption is reversible and EXFAS show that it form inner-sphere surface complexes at high pH [101]. Similarly, removal of Pb from solution containing 810, 1782, 2801 mg/L occur at pH ~5.5 pH and equilibration time of 2 weeks, thus indication high sorption capacity of birnessite for Pb [101]. Also, desorption experiment show that the sorption is reversible at pH ~1 [101]. In another study, sorption of Pb(II), Cu(II), Zn(II), Cd(II) onto hexagonal birnessite was carried out at pH 4.5 and characterised with powder X-ray diffraction (XRD) and X-ray photoelectron spectroscopy (XPS). The result of the study indicates that sorption capacity of biressinte for these metals follow this sequence: $Pb^{2+} \gg Cu^{2+} > Zn^{2+} > Cd^{2+}$ but Pb sorption is up to 3.9 times more than other metals [102].

3.8.3 Magnetite

Synthetic magnetite nano particles was used to remove Hexavalent chromium (Cr(VI)) from synthetic wastewater containing 0 to 50 mg/L Cr(IV) at pH 3.0 and 30°C.Up to 72% of Cr was removed via adsorption and the isotherm fitted to Freundlich model [53]. In addition, rapid sorption of Cu(II), Zn(II), Cd(II), Pb(II) and Ni(II) on synthetic Amine-magnetic Fe-oxide is influenced by pH, ionic strength and the complexation of amino groups [101]. Maximum adsorption occur at pH 6, ionic strength of 1.0 mmol/L NaCl and duration of 60 min and are best described with Pseudo-second-order kinetic model, Langmuir model [101]. The adsorption affinity of the adsorbent follow this sequence Pb(II) > Cu(II) > Zn (II) > Cd(II) > Ni(II) [103].

3.9 Mixture of oxide, clay and/or other materials

Mixture of goethite, humic acid and kaolinite are good adsorbents for lead, cadmium, zinc, nickel and copper and the adsorption is better described with Langmuir and Freundlich isotherm model [104]. However, in a five metal ion system (Quinary), the adsorption of lead, cadmium and nickel are affected negatively by the presence of zinc and copper whereas the presence of lead, cadmium and nickel has synergistic effect on the sorption of zinc and copper [104]. Another study found that low-cost adsorbent Algeria clay that is composed of predominantly montmorillonite and kaolinite has the capacity of adsorbing Cu(II) at pH of 6.5 and 20°C with maximum adsorption capacity of 12.22 mg/g [105]. However, treated Algeria clay under similar condition as the untreated clay has adsorption capacity of 15.40 mg/g [105]. The process of adsorption was spontaneous and exothermic [105]. Sorption of copper (II) ion on palygorskite and sepiolite is enhanced at elevated temperature [106]. The adsorption reactions are endothermically driven but sepiolite sorbs copper spontaneously and has more copper retention capacity compared to palygorskite [106]. In another study, combination of montmorillonite and ZnO show high capacity for adsorption of Pb and Cu from aqueous solution at wide range of pH [107]. The kinetic of the adsorption reaction follow pseudo-second-order whereas the adsorption isotherm is described by Langmuir isotherm [107]. Arsenic adsorption on goethite, amorphous Fe-hydroxide, and Ti(IV)-Fe (III)-Al(III)-pillared bentonite, clay pillared with titanium (IV), iron (III), and aluminium (III) reveal that amorphous Fe-hydroxide has highest adsorption capacity for arsenate and arsenite [108]. Arsenic is stable at pH 7 and is mobilised at pH 4 and 10 but mostly at acidic condition. Mn- oxides, amorphous iron oxides and clay minerals sequester up to 61% of arsenic [109].

Combination of montmorillonite and humic acid in the ratio of 100:3 is efficient for removal of cadmium at pH 8.5 and contact time of 24 h with adsorption capacity of 18.96 mg/g for cadmium [110]. Another study reveal that the adsorption and desorption of cadmium and copper from montmorillonite, allophane, kaolinite, halloysite reveal that sericite has the highest ability for Cd sorption, however, montmorillonite showed greatest retention for Cd [111]. In addition, all clay types has sorption ability for copper with pH of 50% metal sorbed lower than pH of 50% metal sorbed for cadmium sorption [111]. In another study, the removal of Cu and Zn from aqueous solution by Al-montmorillonite, goethite, kaolinite and their mixtures at room temperature and pH 4 reveal that adsorption is via inner and outer sphere complexation [112]. However, mixing of different mineral for sorption of Cu and Zn retards their removal and decreases the exchange of proton and acid/base potential of the reactive sites [112]. The sorption of Zn onto hydrous Mn-oxide (HMO)-coated clay reveals that the affinity the HMO-coated montmorillonite was greater than that of uncoated montmorillonite, and possess linear isotherm at pH 5–6 [113]. X-ray absorption spectroscopy (XAS) reveal reduction of first shell distance at surface loading of 10^{-3} mol and pH 5–7 due to higher electrostatic attraction [113]. In another study, sorption of copper, zinc and lead on soil composed of clay minerals (smectites and vermiculites), carbonates and Fe-oxide show that Copper and lead has higher sorption capacity and retention compared to Zn [114]. Clay minerals adsorbs more metals than other phases, however, for lead, similar capacity was obtained for Fe-oxides [114]. The presence of carbonate in alkaline condition increases the amount of metal uptake, and the mixture of clay minerals and Fe-oxide enhanced adsorption of the metals [114]. Competitive sorption and desorption of Cd, Cr, Cu, Ni, Pb and Zn by iron oxide, Mn oxides, kaolinite, vermiculite and mica from initial solution of 100 mg/L show that kaolinite and mica has strong affinity and retention capability for cd; vermiculite, Cu and Zn; iron oxide and Mn-oxide, Pb [115]. Kaolinite has low retention capability for Cu whereas vermiculite and Mn oxide has greatest retention capability of all the metal [115].

In another study, the effect of increase surface area of clay minerals (kaolinite, montmorillonite and illite) through coating with Fe-oxide, organic matter and Al-oxides for adsorption of heavy metals indicate that coating clay increases the surface area of clay minerals with expectation of Aluminium oxide coated montmorillonite and organic matter coated 2:1 phyllosilicates [116]. Another study found that, both amorphous hydrous manganese oxide (HMO) and HMO-coated montmorillonite sorbs Ni and Pb to form inner-sphere complexes with Ni coordinating to vacant site of Mn-oxide structure and Pb forming bidentate corner-sharing complexes [117]. In addition, another study reveal that montmorillonite clay coated with amorphous (hydrous manganese oxide (HMO), birnessite and pyrolusite has the same surface properties as the coated oxide, however, the surface area of the coated Montmorillonite increases whereas the while the pore size distribution decreased. The HMO- and birnessite-coated clay still retained their pH (point of net zero charge (pnzc)) of 2.8 and 3.1, respectively, [118].

3.10 Advantages of using metal oxides and clays for remediation

The use of clay minerals and oxides is more effective than using other materials for remediation of heavy metals from the environment. From example, the use of sepioloite is better for remediation of Cd compared to $Ca(OH)_2$ [119]. In addition, clay minerals and oxides can be recycled and regenerated for additional remediation use. For example, synthetic palygorskite-Fe oxide exhibited the capability of removing Pb and three cycles reusable potential and retains it magnetic properties for the removal of the heavy metals [85]. Recovery of metals from sorbents can increase the life cycle and long term remediation cost [120]. The estimated zinc that will accumulate in composite substrate in Force Crag Mine is will amount to €7600 in 10 years, however, removal of this substrate after its exhaustion will cost more than €0.8 M whereas recycling the substrate through acid washing is estimated to cost €155,000 [120]. Similarly, oxides and clay minerals can be recycled and reused for remediation of the contaminant. In addition, these sorbents can be used to remediate economic heavy/rare earth metals for commercial and industrial purposes, however, this requires further investigation.

4. Conclusion

Oxides and clay minerals have large surface areas and can sorb heavy metals via adsorption and ion exchange, respectively. Sorption of these toxic metals on oxides and clay minerals reduce their concentration and mobility in aqueous solution and natural environment. This review confirms that sorption is highly dependent on pH, equilibration time, initial concentration of adsorbate, type of adsorbent, temperature, type of adsorbent modification and surface area. In addition, most of the adsorption of ions on clays and oxides are best described by Langmuir models and Pseudo-second-order model. Most remediation techniques employ the use of permeable reactive barriers (PBRs) to react the contaminant from groundwater. These barriers can be Fe-oxides, montmorillonite, Mn oxide-coated montmorillonite, Fe-oxide-coated montmorillonite [121]. The review confirmed that both metal oxides and clay have capability of sequestering heavy metals, however, combination of both metal oxides and modified clay have enhanced capability of removing heavy metals from aqueous solution. These inorganic adsorbents has the regeneration and recycling potentials and can be used to remediate and sequester economic metals for commercial purposes, however, this needs future investigation.

Author details

Ifeoma Mary Ugwu[1*] and Onyedikachi Anthony Igbokwe[2,3]

1 Enugu State University of Science and Technology, Enugu State, Nigeria

2 Ruhr University Bochum, Germany

3 Federal University Ndufe-Alike Ikwo, Ebonyi State, Nigeria

*Address all correspondence to: iu1591@my.bristol.ac.uk

References

[1] Carolin CF, Kumar PS, Saravanan A, Joshiba GJ, Naushad M. Efficient techniques for the removal of toxic heavy metals from aquatic environment: A review. Journal of Environmental Chemical Engineering. 2017;**5**(3): 2782-2799

[2] Lee J-C, Pandey BD. Bio-processing of solid wastes and secondary resources for metal extraction - A review. Waste Management. 2012;**32**(1):3-18

[3] Peligro FR, Pavlovic I, Rojas R, Barriga C. Removal of heavy metals from simulated wastewater by in situ formation of layered double hydroxides. Chemical Engineering Journal. 2016; **306**:1035-1040

[4] Demim S, Drouiche N, Aouabed A, Benayad T, Dendene-Badache O, Semsari SC. Assessment of the physiological effects and heavy metal removal using a response surface approach by *L. gibba*. Ecological Engineering. 2013;**61**:426-435

[5] Filipi M. Mechanisms of cadmium induced genomic instability. Mutation Research/Fundamental and Molecular Mechanisms of Mutagenesis. 2012;**733**: 69-77

[6] Miretzky P, Cirelli AF. Cr(VI) and Cr (III) removal from aqueous solution by raw and modified lignocellulosic materials: a review. Journal Hazardous Materials. 2010;**180**:1-19

[7] Hu J, Chen C, Zhu X, Wang X. Removal of chromium from aqueous solution by using oxidized multiwalled carbon nanotubes. Journal Hazardous Materials. 2009;**162**:1542-1550

[8] Qu X, Alvarez PJJ, Li Q. Applications of nanotechnology in water and wastewater treatment. Water Recourses. 2013;**47**:3931-3946

[9] Cechinel MAP, Ulson De Souza SMAG, Ulson De Souza AA. Study of lead (II) adsorption onto activated carbon originating from cow bone. Journal of Cleaner Production. 2013;**65**: 342-349

[10] Awual MR, Ismael M, Khaleque MA, Yaita T. Ultra-trace copper (II) detection and removal from wastewater using novel meso-adsorbent. Journal of Industrial and Engineering Chemistry. 2015;**20**:2332-2340

[11] Tang WW, Zeng GM, Gong JL, Liang J, Xu P, Zhang C, et al. Impact of humic/fulvic acid on the removal of heavy metals from aqueous solutions using nanomaterials: a review. Science of the Total Environment. 2014;**468–469**:1014-1027

[12] Zhou YT, Branford-White C, Nie HL, Zhu ML. Removal of Cu^{2+} from aqueous solution by chitosan-coated magnetic nanoparticles modified withα-ketoglutaric acid. Journal of Colloid and Interface Science. 2009;**330**:29-37

[13] Petcu C, Purcar V, Radu AL, Ianchis R, Elvira A, Sarbu A, et al. Removal of zinc ions from model wastewater system using bicopolymer membranes with fumed silica. Journal of Water Process Engineering. 2015;**8**:1-10

[14] Kumari S, Chauhan GS. New cellulose-lysine schiff-base-based sensor-adsorbent for mercury ions. ACS Applied Materials & Interfaces. 2014;**6**: 5908-5917

[15] Ahmaruzzaman M. Industrial wastes as low-cost potential adsorbents for the treatment of wastewater laden with heavy metals. Advances in Colloid and Interface Science. 2011;**116**:36-59

[16] lenntech.com. 1998-2018. Available from: http://www.lenntech.com/period

ic/elements/co.htm [Accessed: Jun 10, 2018]

[17] Agency for Toxic Substances and Disease Registry. Toxicological Profile for Vanadium. 2012. Available from: https://www.atsdr.cdc.gov/toxprofiles/tp58.pdf [Accessed: Jun 10, 2018]

[18] Cummings KJ, Virji MA, Park JY, Stanton ML, Edwards NT, Trapnell BC, et al. Respirable indium exposures, plasma indium, and respiratory health among indium-tin oxide (ITO) workers: Respirable indium, plasma indium, and respiratory health. American Journal of Industrial Medicine. 2016;7:59

[19] New Jersey Department of Health and Senior Services. Hazardous Substances Sheet: Indium. 1999. Available from: http://nj.gov/health/eoh/rtkweb/documents/fs/1025.pdf [Accessed: Jun 10, 2018]

[20] New Jersey Department of Health and Senior Services. Hazardous Substances Sheet: Gallium. 2001. Available from: http://nj.gov/health/eoh/rtkweb/documents/fs/0956.pdf [Accessed: Jun 10, 2018]

[21] New Jersey Department of Health and Senior Services. Hazardous Substances Sheet: Molybdenum. 1999. Available from: http://nj.gov/health/eoh/rtkweb/documents/fs/1312.pdf [Accessed: Jun 10, 2018]

[22] Agency for Toxic Substances and Disease Registry. Public Health Statement for Cesium. 2004. Available from: https://www.atsdr.cdc.gov/ToxProfiles/tp157-c1-b.pdf [Accessed: Jun 10, 2018]

[23] Agency for Toxic Substances and Disease Registry. Public Health Statement for Thorium. 1990. Available from: https://www.atsdr.cdc.gov/ToxProfiles/tp147-c1-b.pdf [Accessed: Jun 10, 2018]

[24] Deer WA, Howie RA, Zussman J. An Introduction to the Rock—Forming Minerals. London: Longman; 1992

[25] Uddin MK. A review on the adsorption of heavy metals by clay minerals, with special focus on the past decade. Chemical Engineering Journal. 2017;**308**:438-462

[26] Xu Y, Liang X, Xu Y, Qin X, Huang Q, Wang L, et al. Remediation of heavy metal-polluted agricultural soils using clay minerals: A review. Pedosphere. 2017;**27**(2):193-204

[27] Fu FL, Wang Q. Removal of heavy metal ions from wastewaters: A review. Journal of Environmental Management. 2011;**92**(3):407-418

[28] Wang XL, Li Y. Measurement of Cu and Zn adsorption onto surficial sediment components: New evidence for less importance of clay minerals. Journal of Hazardous Materials. 2011; **189**(3):719-723

[29] Koschinsky A, Winkler A, Fritsche U. Importance of different types of marine particles for the scavenging of heavy metals in the deep-sea bottom water. Applied Geochemistry. 2003;**18** (5):693-710

[30] Alby D, Charnay C, Heran M, Prelot B, Zajac J. Recent developments in nanostructured inorganic materials for sorption of cesium and strontium: Synthesis and shaping, sorption capacity, mechanisms, and selectivity-A review. Journal of Hazardous Materials. 2018;**344**:511-530

[31] Raval NP, Shah PU, Shah NK. Adsorptive removal of nickel (II) ions from aqueous environment: A review. Journal of Environmental Management. 2016;**179**:1-20

[32] Mahar A, Wang P, Li RH, Zhang ZQ. Immobilization of lead and cadmium in contaminated soil using

amendments: A review. Pedosphere. 2015;**25**(4):555-568

[33] De Gisi S, Lofrano G, Grassi M, Notarnicola M. Characteristics and adsorption capacities of low-cost sorbents for wastewater treatment: A review. Sustainable Materials and Technologies. 2016;**9**:10-40

[34] Zhao G, Wu X, Tan X, Wang X. Sorption of heavy metal ions from aqueous solutions: A review. The Open Colloid Science Journal. Bentham Open; 2011;**4**:19-31. DOI: 1876-5300/11

[35] Katz LE, Hayes KF. Surface complexation modeling .2. Strategy for modeling polymer and precipitation reactions at high surface coverage. Journal of Colloid and Interface Science. 1995;**170**(2):491-501

[36] Fukushi K. Modelling sorption processes of trace elements by earth surface materials. Journal of Geography-Chigaku Zasshi. 2017;**126**(3):325-341

[37] Ugwu IM, Sherman DM. Irreversibility of sorption of cobalt to goethite (alpha-FeOOH) and disparities in dissolution of aged synthetic Co-goethite. Chemical Geology. 2017;**467**: 168-176

[38] Stumm W. Chemistry of Solid-Water Interface. New York: Wiley; 1992

[39] Langmuir D. Aqueous Environmental Geochemistry. London: Prentice Hal; 1997

[40] Appelo CAJ, Geochemistry PD. Groundwater and Pollution. Rotterdam: Balkema; 2005

[41] Davis JA, Kent DB. Surface complexation modeling in aqueous geochemistry. Mineral-Water Interface Geochemistry. 1990;**23**:177-260

[42] Sherman DM. Surface complexation modelling: Mineral-fluid equilibria at the molecular scale. Reviews in Mineralogy and Geochemistry. 2009;70: 181-205

[43] White WM. Geochemistry. UK: Willey-Blackwell; 2013. 668 pp

[44] Goldberg S. Adsorption Models Incorporated into Chemical Equilibrium Models. Chemical Equilibrium and Reaction Models. USA; 1995

[45] Herbelin A, Westall J. A Computer Program for Determination of Chemical Equilibrium Constants From Experimental Data. 3.2 ed. Corvallis, OR: Department of Chemistry, Oregon State University; 1996

[46] Sherman DM, Peacock CL, Hubbard CG. Surface complexation of U(VI) on goethite (alpha-FeOOH). Geochimica Et Cosmochimica Acta. 2008;**72**(2):298-310

[47] Ijagbemi CO, Baek M-H, Kim D-S. Montmorillonite surface properties and sorption characteristics for heavy metal removal from aqueous solutions. Journal of Hazardous Materials. 2009;**166**(1): 538-546

[48] Dumat C, Cheshire MV, Fraser AR, Shand CA, Staunton S. The effect of removal of soil organic matter and iron on the adsorption of radiocaesium. European Journal of Soil Science. 1997; **48**(4):675-683

[49] Khalili FI, Salameh NH, Shaybe MM. Sorption of uranium (VI) and thorium (IV) by Jordanian Bentonite. Journal of Chemistry. 2013;**13**:13

[50] Ararem A, Bouras O, Arbaoui F. Adsorption of caesium from aqueous solution on binary mixture of iron pillared layered montmorillonite and goethite. Chemical Engineering Journal. 2011;**172**(1):230-236

[51] Karamanis D, Assimakopoulos PA. Efficiency of aluminum-pillared montmorillonite on the removal of cesium and copper from aqueous solutions. Water Research. 2007;**41**(9): 1897-1906

[52] Zheng X, Dou J, Yuan J, Qin W, Hong X, Ding A. Removal of Cs^+ from water and soil by ammonium-pillared montmorillonite/Fe_3O_4 composite. Journal of Environmental Sciences. 2017;**56**:12-24

[53] Padmavathy KS, Madhu G, Haseena PV. A study on effects of pH, adsorbent dosage, time, initial concentration and adsorption isotherm study for the removal of hexavalent chromium (Cr (VI)) from wastewater by magnetite nanoparticles. In: International Conference on Emerging Trends in Engineering, Science and Technology (Icetest—2015), Procedia Technology; Vol. 24. 2016. pp. 585-594

[54] Yoon TH, Johnson SB, Musgrave CB, Brown GE. Adsorption of organic matter at mineral/water interfaces: I. ATR-FTIR spectroscopic and quantum chemical study of oxalate adsorbed at boehmite/water and corundum/water interfaces. Geochimica et Cosmochimica Acta. 2004;**68**(22):4505-4518

[55] Peacock CL, Sherman DM. Sorption of Ni by birnessite: Equilibrium controls on Ni in seawater. Chemical Geology. 2007;**238**(1–2):94-106

[56] Peacock CL, Sherman DM. Copper (II) sorption onto goethite, hematite, and lepidocrocite: A surface complexation model based on ab initio molecular geometries and EXAFS spectroscopy (Vol 68, p. 2623, 2004). Geochimica et Cosmochimica Acta. 2005;**69**(21):5141-5142

[57] Kwon KD, Refson K, Sposito G. Understanding the trends in transition metal sorption by vacancy sites in birnessite. Geochimica et Cosmochimica Acta. 2013;**101**:222-232

[58] Yang S, Ren X, Zhao G, Shi W, Montavon G, Grambow B, et al. Competitive sorption and selective sequence of Cu(II) and Ni(II) on montmorillonite: Batch, modeling, EPR and XAS studies. Geochimica et Cosmochimica Acta. 2015;**166**:129-145

[59] Khan TA, Singh VV. Removal of cadmium (II), lead (II), and chromium (VI) ions from aqueous solution using clay. Toxicological and Environmental Chemistry. 2010;**92**(8):1435-1446

[60] Eloussaief M, Benzina M. Efficiency of natural and acid-activated clays in the removal of Pb(II) from aqueous solutions. Journal of Hazardous Materials. 2010;**178**(1–3):753-757

[61] Demirkiran AR, Acemioglu B, Gonen T. Sorption of copper and nickel ions from solution by clay minerals. Oxidation Communications. 2016;**39**(1): 817-829

[62] Akpomie KG, Dawodu FA, Adebowale KO. Mechanism on the sorption of heavy metals from binary-solution by a low cost montmorillonite and its desorption potential. Alexandria Engineering Journal. 2015;**54**(3):757-767

[63] Akpomie KG, Dawodu FA. Acid-modified montmorillonite for sorption of heavy metals from automobile effluent. Beni-Suef University Journal of Basic and Applied Sciences. 2016;**5**:1-12

[64] de Pablo L, Lourdes Chavez M, Abatal M. Adsorption of heavy metals in acid to alkaline environments by montmorillonite and Ca-montmorillonite. Chemical Engineering Journal. 2011;**171**(3):1276-1286

[65] Kowal-Fouchard A, Drot R, Simoni E, Marmier N, Fromage F, Ehrhardt JJ. Structural identification of europium

(III) adsorption complexes on montmorillonite. New Journal of Chemistry. 2004;**28**(7):864-869

[66] Chen C, Liu H, Chen T, Chen D, Frost RL. An insight into the removal of Pb(II), Cu(II), Co(II), Cd(II), Zn(II), Ag(I), Hg(I), Cr(VI) by Na(I)-montmorillonite and Ca(II)-montmorillonite. Applied Clay Science. 2015;**118**:239-247

[67] Sen Gupta S, Bhattacharyya KG. Adsorption of heavy metals on kaolinite and montmorillonite: A review. Physical Chemistry Chemical Physics. 2012;**14**(19):6698-6723

[68] Rao RAK, Kashifuddin M. Adsorption studies of Cd(II) on ball clay: Comparison with other natural clays. Arabian Journal of Chemistry. 2016;**9**:S1233-S1S41

[69] Jiang MQ, Jin XY, Lu XQ, Chen ZL. Adsorption of Pb(II), Cd(II), Ni(II) and Cu(II) onto natural kaolinite clay. Desalination. 2010;**252**(1–3):33-39

[70] Bhattacharyya KG, Sen Gupta S. Adsorptive accumulation of Cd(II), Co(II), Cu(II), Pb(II) and Ni(II) ions from water onto kaolinite: Influence of acid activation. Adsorption Science & Technology. 2009;**27**(1):47-68

[71] Bhattacharyya KG, Gupta SS. Adsorption of chromium (VI) from water by clays. Industrial and Engineering Chemistry Research. 2006;**45**(45):7232-7240

[72] Giannakopoulou F, Haidouti C, Chronopoulou A, Gasparatos D. Sorption behavior of cesium on various soils under different pH levels. Journal of Hazardous Materials. 2007;**149**:553-556

[73] Cherif MA, Martin-Garin A, Gerard F, Bildstein O. A robust and parsimonious model for caesium sorption on clay minerals and natural clay materials. Applied Geochemistry. 2017;**87**:22-37

[74] Martin LA, Wissocq A, Benedetti MF, Latrille C. Thallium (Tl) sorption onto illite and smectite: Implications for Tl mobility in the environment. Geochimica et Cosmochimica Acta. 2018;**230**:1-16

[75] Freitas ED, Carmo ACR, Almeida Neto AF, Vieira MGA. Binary adsorption of silver and copper on Verde-lodo bentonite: Kinetic and equilibrium study. Applied Clay Science. 2017;**137**:69-76

[76] Calagui MJC, Senoro DB, Kan C-C, Salvacion JWL, Futalan CM, Wan M-W. Adsorption of indium (III) ions from aqueous solution using chitosan-coated bentonite beads. Journal of Hazardous Materials. 2014;**277**:120-126

[77] Yin Z, Pan D, Liu P, Wu H, Li Z, Wu W. Sorption behavior of thorium (IV) onto activated bentonite. Journal of Radioanalytical and Nuclear Chemistry. 2018;**316**(1):301-312

[78] Dutta J, Mishra AK. Influence of the presence of heavy metals on the behaviour of bentonites. Environmental Earth Sciences. 2016;**75**(11)

[79] Lukman S, Essa MH, Nuthu D, Mu'azu N, Bukhari A, Basheer C. Adsorption and desorption of heavy metals onto natural clay material: Influence of initial pH. Journal of Environmental Science and Technology. 2013;**6**:1-15

[80] Sipos P, Kis VK, Baldzs R, Toth A, Kovacs I, Nemeth T. Contribution of individual pure or mixed-phase mineral particles to metal sorption in soils. Geoderma. 2018;**324**:1-8

[81] Sdiri A, Higashi T, Chaabouni R, Jamoussi F. Competitive removal of heavy metals from aqueous solutions by montmorillonitic and calcareous clays.

Water Air and Soil Pollution. 2012;**223**(3):1191-1204

[82] Sharma YC. Thermodynamics of removal of cadmium by adsorption on an indigenous clay. Chemical Engineering Journal. 2008;**145**(1):64-68

[83] Weng C-H, Sharma YC, Chu S-H. Adsorption of Cr(VI) from aqueous solutions by spent activated clay. Journal of Hazardous Materials. 2008;**155**(1–2):65-75

[84] Yang F, Sun S, Chen X, Chang Y, Zha F, Lei Z. Mg-Al layered double hydroxides modified clay adsorbents for efficient removal of Pb^{2+}, Cu^{2+} and Ni^{2+} from water. Applied Clay Science. 2016;**123**:134-140

[85] Rusmin R, Sarkar B, Tsuzuki T, Kawashima N, Naidu R. Removal of lead from aqueous solution using superparamagnetic palygorskite nanocomposite: Material characterization and regeneratilon studies. Chemosphere. 2017;**186**:1006-1015

[86] Medhi H, Bhattacharyya KG. Kinetic and mechanistic studies on adsorption of Cu(II) in aqueous medium onto montmorillonite K10 and its modified derivative. New Journal of Chemistry. 2017;**41**(22):13533-13552

[87] Lee SM, Lalhmunsiama T, Tiwari D. Porous hybrid materials in the remediation of water contaminated with As(III) and As(V). Chemical Engineering Journal. 2015;**270**:496-507

[88] Franco F, Benitez-Guerrero M, Gonzalez-Trivino I, Perez-Recuerda R, Assiego C, Cifuentes-Melchor J, et al. Low-cost aluminum and iron oxides supported on dioctahedral and trioctahedral smectites: A comparative study of the effectiveness on the heavy metal adsorption from water. Applied Clay Science. 2016;**119**:321-332

[89] Eloussaief M, Sdiri A, Benzina M. Modelling the adsorption of mercury onto natural and aluminium pillared clays. Environmental Science and Pollution Research. 2013;**20**:469-479

[90] Timofeev KL, Maltsev GI, Sviridov AV. Sorption kinetics of indium, iron, and zinc ions on modified montmorillonite. Moscow University Chemistry Bulletin. 2017;**72**(3):128-134

[91] Alemayehu DD, Sk S, Tessema DA. Assessment of the adsorption capacities of fired clay soils from Jimma (Ethiopia) for the removal of Cr(VI) from aqueous solution. Universal Journal of Environmental Research and Technology. 2012;**2**:411-420

[92] Cornell RMSU. The Iron Oxides: Structure, Properties, Reactions, Occurances and Uses. Germany: John Wiley and Sons; 2003

[93] Yuan W, Saldi GD, Chen J, Zuccolini MV, Birck J-L, Liu Y, et al. Gallium isotope fractionation during Ga adsorption on calcite and goethite. Geochimica et Cosmochimica Acta. 2018;**223**:350-363

[94] Peacock C, Sherman D. Vanadium (V) adsorption onto goethite (alpha-FeOOH) at p. 1.5 to 12: A surface complexation model based on ab initio molecular geometries and EXAFS spectroscopy. Geochimica et Cosmochimica Acta. 2004;**68**:1723-1733

[95] Fischer L, Bruemmer GW, Barrow NJ. Observations and modelling of the reactions of 10 metals with goethite: Adsorption and diffusion processes. European Journal of Soil Science. 2007;**58**(6):1304-1315

[96] Trivedi P, Axe L, Dyer J. Adsorption of metal ions onto goethite: single-adsorbate and competitive systems. Colloids and Surfaces A—

Physicochemical and Engineering Aspects. 2001;**191**(1–2):107-121

[97] White SJO, Hussain FA, Hemond HF, Sacco SA, Shine JP, Runkel RL, et al. The precipitation of indium at elevated pH in a stream influenced by acid mine drainage. Science of the Total Environment. 2017;**574**:1484-1491

[98] Ugwu IM. Sorption of Ni and Co on goethite with application to Ni laterite formation. United Kingdom: University of Bristol; 2015

[99] Lefkowitz JP, Elzinga EJ. Structural alteration of hexagonal birnessite by aqueous Mn(II): Impacts on Ni(II) sorption. Chemical Geology Journal. 2017;**466**:524-532

[100] Sherman DM, Peacock CL. Surface complexation of Cu on birnessite (delta-MnO_2): Controls on Cu in the deep ocean. Geochimica et Cosmochimica Acta. 2010;**74**(23):6721-6730

[101] Bacon CGD. Surface complexation of Pb and Zn onto birnessite (δ-MnO_2); controls on pollution in soils and groundwater. United Kingdom: University of Bristol; 2014

[102] Wang Y, Feng X, Villalobos M, Tan W, Liu F. Sorption behavior of heavy metals on birnessite: Relationship with its Mn average oxidation state and implications for types of sorption sites. Chemical Geology Journal. 2012;**292-293**:25-34

[103] Lin S, Liu LL, Yang Y, Lin KF. Study on preferential adsorption of cationic-style heavy metals using amine-functionalized magnetic iron oxide nanoparticles (MIONPs-NH_2) as efficient adsorbents. Applied Surface Science. 2017;**407**:29-35

[104] Unuabonah EI, Olu-Owolabi BI, Adebowale KO. Competitive adsorption of metal ions onto goethite-humic acid-modified kaolinite clay. International Journal of Environmental Science and Technology. 2016;**13**(4):1043-1054

[105] Djebbar M, Djafri F. Adsorption of Cu(II) on natural and treated clays. Water Quality Research Journal of Canada. 2016;**51**(1):26-32

[106] Khosravi P, Shirvani M, Bakhtiary S, Shariatmadari H. Energetic and entropic features of Cu(II) sorption equilibria on fibrous clay minerals. Water Air and Soil Pollution. 2016:227-354

[107] Sani HA, Ahmad MB, Hussein MZ, Ibrahim NA, Musa A, Saleh TA. Nanocomposite of ZnO with montmorillonite for removal of lead and copper ions from aqueous solutions. Process Safety and Environmental Protection. 2017;**109**:97-105

[108] Lenoble V, Bouras O, Deluchat V, Serpaud B, Bollinger JC. Arsenic adsorption onto pillared clays and iron oxides. Journal of Colloid and Interface Science. 2002;**255**(1):52-58

[109] Shaw D. Mobility of arsenic in saturated, laboratory test sediments under varying pH conditions. Engineering Geology. 2006;**85**(1–2):158-164

[110] Zeng X, Wang Z, Ji Z, Wei S. A new montmorillonite/humic acid complex prepared in alkaline condition to remove cadmium in waste water. Polish Journal of Environmental Studies. 2015;**24**(2):817-821

[111] Itami K, Yanai J. Sorption and desorption properties of cadmium and copper on soil clays in relation to charge characteristics. Soil Science and Plant Nutrition. 2006;**52**(1):5-12

[112] Egirani DE, Baker AR, Andrews JE. Copper and zinc removal from aqueous solution by mixed mineral systems—II.

The role of solution composition and aging. Journal of Colloid and Interface Science. 2005;**291**(2):326-333

[113] Boonfueng T, Axe L, Xu Y, Tyson TA. The impact of Mn oxide coatings on Zn distribution. Journal of Colloid and Interface Science. 2006;**298**(2):615-623

[114] Sipos P, Nemeth T, Kis VK, Mohai I. Sorption of copper, zinc and lead on soil mineral phases. Chemosphere. 2008;**73**(4):461-469

[115] Vega FA, Covelo EF, Vazquez JJ, Andrade L. Influence of mineral and organic components on copper, lead, and zinc sorption by acid soils. Journal of Environmental Science and Health Part A—Toxic/Hazardous Substances & Environmental Engineering. 2007;**42**(14):2167-2173

[116] Zhuang J, Yu GR. Effects of surface coatings on electrochemical properties and contaminant sorption of clay minerals. Chemosphere. 2002;**49**(6):619-628

[117] Boonfueng T, Axe L, Xu Y, Tyson TA. Nickel and lead sequestration in manganese oxide-coated montmorillonite. Journal of Colloid and Interface Science. 2006;**303**(1):87-98

[118] Boonfueng T, Axe L, Xu Y. Properties and structure of manganese oxide-coated clay. Journal of Colloid and Interface Science. 2005;**281**(1):80-92

[119] Zhu Q-H, Huang D-Y, Zhu G-X, Ge T-D, Liu G-S, Zhu H-H, et al. Sepiolite is recommended for the remediation of Cd-contaminated paddy soil. Acta Agriculturae Scandinavica Section B-Soil and Plant Science. 2010;**60**(2):110-116

[120] Bailey MT, Gandy CJ, Jarvis AP. Reducing life-cycle costs of passive mine water treatment by recovery of metals from treatment wastes. In: Drebenstedt C, Paul M, editors. Proceedings IMWA 2016, Freiberg/Germany, Mining Meets Water – Conflicts and Solutions. 2016

[121] Park Y, Shin WS, Choi S-J. Sorptive removal of cobalt, strontium and cesium onto manganese and iron oxide-coated montmorillonite from groundwater. Journal of Radioanalytical and Nuclear Chemistry. 2012;**292**(2):837-852

Chapter 8

Marine Algae Bioadsorbents for Adsorptive Removal of Heavy Metals

Mazen K. Nazal

Abstract

With the shortage of freshwater resources and as wastewater output of huge industries as well as pollution that might be happening in the ecosystem, wastewater treatment is of utmost importance. Removal of pollutants such as heavy metals from wastewater would provide an exceptional alternative water resource. Extensive research has been done to develop an operative technology to overcome the toxicity and the negative environmental impact of heavy metals and their ionic forms. In this book chapter, biomass bioadsorbents utilizing marine algae for adsorptive removal of heavy metal pollutants from wastewater were discussed. The most common adsorption isotherms and kinetic models, which used to study their nature of adsorption, were also covered.

Keywords: macroalgae, adsorption isotherm, adsorption kinetics, wastewater, adsorbents

1. Introduction

1.1 Wastewater

Water which is the key element responsible for life in the world is becoming more valuable due to the increased consumption and demand. In order to provide a locally controlled water supply, wastewater recycling offers great environmental advantages. Recycling of water can corporate in decreasing the consumption of water from sensitive ecosystem, reducing the environmental pollution, and even preventing accumulation of pollutants in our ecosystem. The US Environmental Protection Agency (USEPA) has suggested three stages of water recycling; in the primary stage that can be achieved by a sedimentation process, normally the produced water is not suitable for any use. The biological oxidation and disinfection process are used to reach the secondary stage. The produced water from that stage can be used mainly for irrigation of nonfood crop and industrial cooling system. The tertiary stage in wastewater treatment is reached using chemical, coagulation, filtration, and disinfection processes. Produced water in the tertiary stage can be employed mostly for irrigation of food crops and landscape, washing of vehicles, and flushing toilet [1]. Good quality water (i.e., water free of contaminants) is essential to human health and a critical feedstock in a variety of key industries including oil and gas, petrochemicals, pharmaceuticals, and food. The available supplies of water are decreasing due to (1) low precipitation, (2) increased population growth, (3) more strict health-based regulations, and (4) competing demands from a variety of users, e.g., industrial, agricultural, and urban development. In addition, our water today became such type of cocktail of chemicals that has more than 100 of toxic compounds, viruses, bacteria, and metals. Consequently, water scientists and engineers are seeking alternative sources of water and new technologies for wastewater treatment and recycling. These wastewaters include but not limited to sewage effluent, contaminated surface or groundwater, and industrial wastewater. Water recovery-recycle-reuse has proven to be effective and successful in creating a new and reliable water supply while not compromising public health [2].

1.2 Heavy metals

Water pollution with contaminants became a global issue. Among of these contaminants, heavy metals have a greater concern mainly due to their bioaccumulation, toxicity, and non-biodegradability. Their non-biodegradability nature makes their existence in water to cause great risk to living organisms. Accordingly, many government environmental agencies such the US Environmental Protection Agency (USEPA) and World Health Organization (WHO) have set the maximum acceptable concentration level for heavy metals in recycled water. Therefore, different methodologies, with varying level of success, have been employed to remove these contaminations from water and wastewater. Biological treatment (aerobic and anaerobic), coagulation, precipitation, oxidation, membrane, and filtration are common methods of removing microorganisms and ionic and cationic compounds from wastewater streams. The performance of these methods is generally acceptable at low concentration of heavy metals below few hundred ppm, which is the main drawback of them. Even though most of the wastewater treatment technologies available today are effective, they are often costly and time-consuming methods. Bioadsorption is considered as among the most promising low-cost process for wastewater treatment. Numerous materials were used as adsorbents to remove heavy metal ions from water, such as metal oxides, activated carbon, zeolite, chitin, metal sulfide, resin, etc. The search for new and more effective materials to be used as bioadsorbent materials has a continuous effort and been considered by many researchers. Since 1990 till now, there are more than 5000 publications in the field of bioadsorption of heavy metals, and approximately 6% of these publications have been concerned on using marine algae [3]. **Figure 1**(**a** and **b**) shows the dramatic increase in both the number of publications and their citations versus time.

1.3 Marine algae

Marine algae are one of the most highly available natural resources in tropical ecosystem where around 2 million tons of them are collected from seas and oceans and cultured in artificial system [4]. They are useful in different applications such as pharmaceutical, food, and cosmetic industries. Algae have rich biochemical composition; therefore, its biomass is a promising material to be used as bioadsorbent to decontaminate water and wastewater by removing pollutants such as heavy metals [5, 6]. Marine algae are commonly known as seaweeds, and they had a great potential to be used in pollutant removal process as a promising bioadsorbents material. This is due to their renewable availability, distinct properties, and high biosorption capacity. Seaweeds are divided into three main broad groups, namely, (i) green (Chlorophyta), (ii) red (Rhodophyta), and (iii) brown (Phaeophyta) algae. Marine algae have many advantages for bioadsorption. Among them brown algae provided the best adsorption capacities due to their cell wall structure and components.

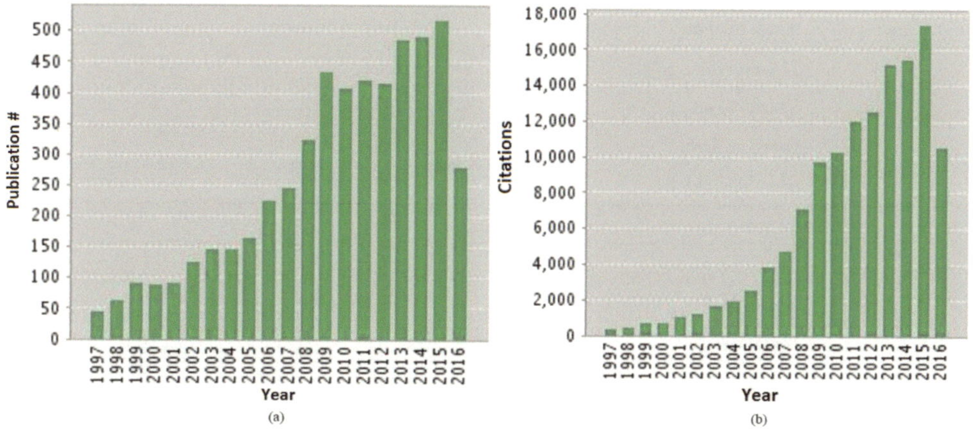

Figure 1.
Histograms for (a) number of publications in the field of biosorption of heavy metals and (b) the number of citations each year on these publication [3].

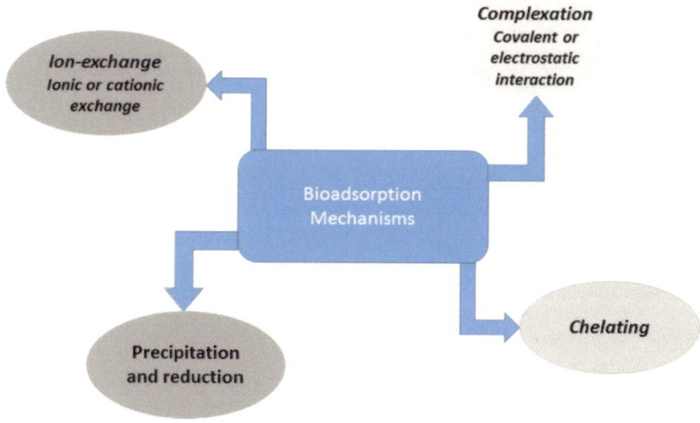

Figure 2.
Classification of metal uptake mechanism by bioadsorbents.

The cell wall of brown algae has a lot of active chemical functional groups such hydroxyl, carboxylic acid, amine, imidazole, phosphate, phenolic, thioether, and sulfhydryl which offer a selective binding and interaction with metals and pollutants in the bioadsorption process. It contains mainly cellulose, a group of salts of sodium, potassium magnesium, and calcium, and alginate, which is a type of polysaccharide (anionic copolymer) [7].

Figure 2 illustrates the main four mechanisms of heavy metal uptake by bioadsorbents. The first one is ion-exchange process including ionic or cationic exchange. The surface of the cell wall contains mainly organic nitrogen group in the case of ionic exchange or hydroxyl and organic sulfate or phosphate in the case of cationic exchange. The uptake mechanism can be a complexation through a covalent or electrostatic interaction where the metal ions form a complex compound with organic molecules. The third mechanism is chelation which involves an interaction between the metal and an organic compound that has more than one electron donor functional group. The last one is through precipitation that occurs when the pH of the solution varies due to cellular metabolism or when the concentration of metals increases [8].

Table 1 summarizes some of the marine algae (red, green, and brown), those used for removal of transition, actinide, or lanthanide metals. Many researchers found that the *Sargassum* brown algae has a high adsorption capacity

Number	Name of algae	Removed metals	Ref.
1	*Sargassum* sp.	Cu	[9]
2	*Sargassum* sp.	Sm and Pr	[10]
3	*Spirogyra* spp.	Cr	[11]
4	*Sargassum vulgaris*	Cd and Ni	[12]
5	*Sargassum hystrix*	Pb	[13]
6	*Sargassum natans*	Pb	[13]
7	*Sargassum hemiphyllum*	Ni and Cu	[14]
8	*Sargassum wightii*	Ni	[15]
9	*Sargassum* sp.	Cr	[15]
10	*Sargassum honeri and S. hemiphyllum*	Ho, Dy, Lu, and Yb	[16]
11	*Sargassum ilicifolium*	Ni and Co	[17]
12	*Sargassum* sp.	La, Nb, Eu, and Gd	[18]
13	*Sargassum muticum* and *Fucus spiralis* (brown algae)	Pd, Zn, and Cd	[19]
14	*Fucus vesiculosus* (brown algae)	Cu	[20]
15	*Palmaria palmata* (red algae)	Cu	[20]
16	*Fucus spiralis* (brown algae)	Cu	[20]
17	*Ulva* sp. (green algae)	Cu	[20]
18	*Fucus ceranoides* and *Fucus serratus* (brown algae)	Cd	[21]
19	*Laminaria japonica*	Cd, Pb, and Fe	[22]
20	*Laminaria japonica* (washed or oxidized by potassium permanganate)	Pb	[23]
21	*Gracilaria fischeri*	Cu and Cd	[24]
22	*Gracilaria* sp.	Cd, Cu, Zn, Pb, and Ni	[25]
	Padina sp.	Cd, Cu, Zn, Pb, and Ni	[25]
23	*Pilayella littoralis*	Cr, Fe, Al, Cd, Cu, Zn, Co, and Ni	[26]
24	*Cladophora crispata*	Pb, Cu, Cd, and Ag	[27]
25	*Cladophora fascicularis*	Cu and Pb	[28]
26	*Ecklonia* sp.	Cr	[29]
27	*Colpomenia sinuosa*	Ni and Cu	[14]
28	*Petalonia fascia*	Ni and Cu	[14]
29	*Ulva fascia*	Ni and Cu	[14]
30	*Padina pavonica*	Ni and Cd	[12]
31	*Sargassum cymosum*	Cr	[30]
32	*Turbinaria conoides*	Pb	[31]
33	*Laurencia obtusa*	Cd, Co, Cr, Cu, and Ni	[32]
34	*Ulva reticulata*	Zn	[33]
35	*Ascophyllum nodosum* *Fucus spiralis* *Laminaria hyperborean* *Pelvetia canaliculata*	Cu, Ni, Zn, and Ca	[34]

Table 1.
Marine algae used in bioadsorption removal of heavy and lanthanide metals.

to remove heavy metals such as Cu, Ni, Cd, Pd, Cr, Sm, and Pr from their solution efficiently due to its cell wall structure that is rich in active bioadsorption sites [9, 10, 15, 17–19]. Mostly, bioadsorption offers many advantages over the bioaccumulation process since bioadsorbents are available commonly as by-product or waste, as well as they do not need growth media and growth conditions. As a result, they are considered low-cost materials with high possibility to be reused for many cycles. The literatures show that marine algae can be used for the removal of heavy metals in dead or live forms. However, in industrial applications, the nonliving marine algae provide more practical bioadsorbent materials for the removal of pollutants. This is because toxicity of heavy metals and other pollutants do not affect dead biomass. In addition, the performance of those bioadsorbents can be improved by physical treatments such as heating or chemical processing such as acid or base treatments. This enhancement in their biosorption capacity is attributed to activation of the adsorption sites as well as rearrangement of the cell wall structure to be more accessible and compatible for pollutants capturing and removal [35].

2. The nature and kinetics of bioadsorption

2.1 Adsorption isotherm models

An idea about the adsorption process is predicted using the correlation between the pressure or the concentration of adsorbate and the adsorption capacity (X/m) at constant temperature as shown in **Figure 3**.

The amount of adsorbate (X) adsorbed should be normalized by the mass of adsorbent (m) to allow comparison of different materials. From **Figure 1**, it can be predicted that after the saturation point, the number of adsorption sites on the adsorbent is occupied, and the vacancies became limited so that the adsorption does not occur anymore. There are five general types of adsorption isotherms. They are as follows:

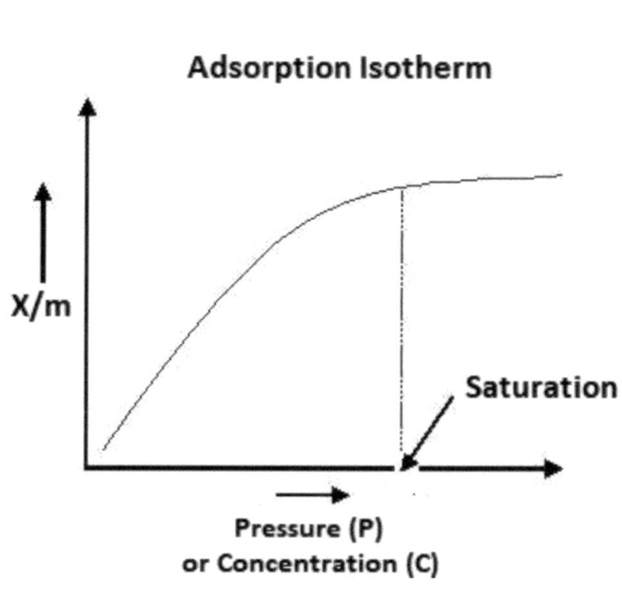

Figure 3.
Adsorption isotherm.

- Type I adsorption isotherm (shown in **Figure 2**)

The main characteristics of this type are (i) there is a monolayer adsorption and (ii) it might be explained using the Langmuir adsorption isotherm.

- Type II adsorption isotherm

Figure 4 shows a typical adsorption isotherm curve of type II. This type of adsorption shows a large deviation from the Langmuir isotherm model and a flat region, which is corresponding to a monolayer formation.

- Type III adsorption isotherm

This type of isotherm indicates that there is no flat region as shown in **Figure 5**, and also there are formations of multilayer adsorption.

- Type IV adsorption isotherm

It can be depicted from **Figure 6** that there is a monolayer formation (intermediate region), which is followed by a multilayer formation at certain adsorbate concentration. At low concentration of adsorbate, the adsorption is mostly similar to type II adsorption isotherm.

- Type V adsorption isotherm

It is similar to type IV with a difference in the range of adsorbate's concentration where the monolayer and multilayer start the formation as shown in **Figure 7**.

The adsorption isotherms usually are being studied to understand the adsorption behavior modulation and to calculate the adsorption capacity for the adsorbents, so the data analysis is done using a linear/nonlinear least squares methods of adsorption isotherms, where they describe the relationship between the adsorbed amount of adsorbate and its equilibrium concentration in the solution.

The Freundlich, Langmuir, Temkin, Sips, and Redlich-Peterson models are the most common types of the adsorption isotherms to describe the metal ion bioadsorption from their single component solution.

The Freundlich isotherm (Eq. 1) is an empirical model where the adsorption occurs on heterogeneous adsorption sites on adsorbent surface, which is the general case in macroalgae bioadsorbents:

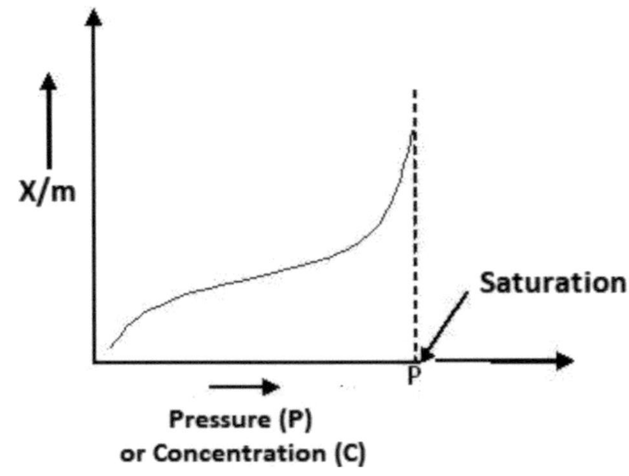

Figure 4.
Type II adsorption isotherm.

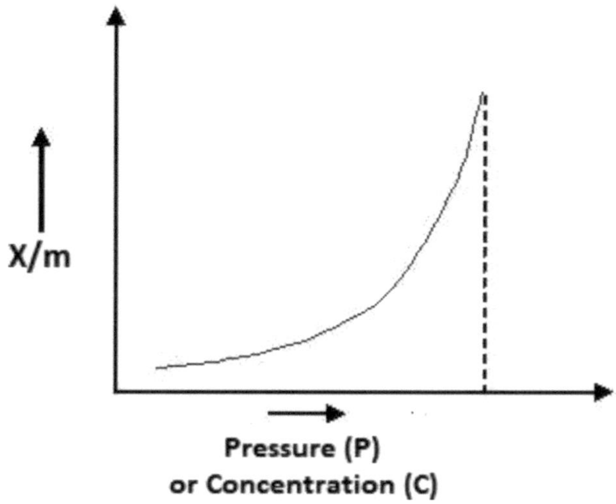

Figure 5.
Type III adsorption isotherm.

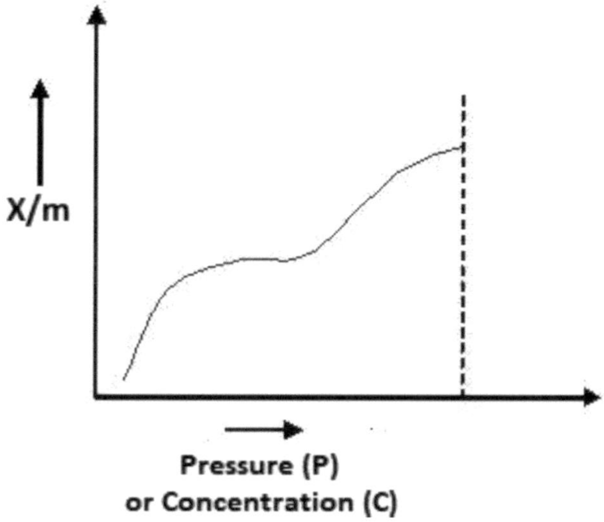

Figure 6.
Type IV adsorption isotherm.

$$q_e = K_f C_e^{1/n} \quad (1)$$

where

q_e: The adsorption density at equilibrium (mg adsorbate/g of adsorbent).

C_e: The residual adsorbate concentration in the solution (mg/L) at equilibrium.

K_f: The relative adsorption capacity ($mg^{1-1/n}1^{1/n}/g$).

n: The unit less constants reflect the adsorption intensity.

A plot of $\ln C_e$ against $\ln q_e$ will give a straight line with a slope 1/n and intercept $LnK_{f.}$ Smaller 1/n greater expected heterogeneity [35]. It is worthy here to note that usually the adsorption data have a good fit with the Freundlich isotherm model due to the well-known insensitivity of its linear form (ln-ln plot).

The Langmuir adsorption isotherms model is considered as the best known for describing a monolayer chemical adsorption process on homogenous adsorption

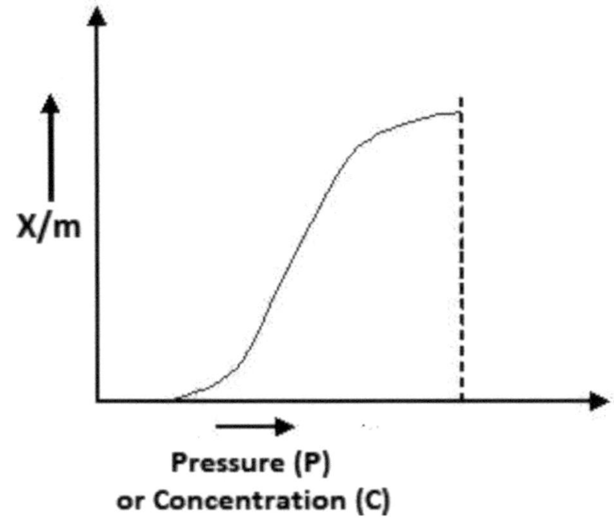

Figure 7.
Type V adsorption isotherm.

sites on adsorbent surfaces. It partially considers the thermodynamic in the adsorption process. It is expressed in Eq. (2):

$$q_e = \frac{q_{max} b C_e}{1 + b C_e} \tag{2}$$

where

q_e: The adsorption capacity at equilibrium (mg of adsorbate/g of adsorbent).

C_e: The residual adsorbate concentration at equilibrium in solution (mg/L).

q_{max}: The maximum adsorption capacity corresponding to monolayer coverage (mg of analyte adsorbed/g of adsorbent).

b: The Langmuir constant correlated to the adsorption energy (1/mg adsorbate).

The essential features of the Langmuir isotherm may be expressed in terms of equilibrium parameter R_L (Eq. 3), which is a dimensionless constant referred to as separation factor or equilibrium parameter [36]:

$$R_L = \frac{1}{1 + (1 + K_L C_e)} \tag{3}$$

The most used linear form of the Langmuir model is the following form (Eq. 4), which is also called reciprocal Langmuir plot:

$$\frac{C_e}{q_e} = \frac{1}{q_{max} b} + \frac{C_e}{q_{max}} \tag{4}$$

Plotting C_e/q_e versus C_e from the experimental data gives a linear regression where the slope for that plot gives the experimental maximum adsorption capacity q_{max}, and the intercept gives the Langmuir constant b.

There are another three linear transformation forms of the Langmuir isotherm models: (1) the distribution coefficient or Scatchard plot, (2) Eadie-Hofstee plot, and (3) double reciprocal Lineweaver-Burk plot. Every one of these four linear transformation forms gives a greater weighing to low adsorption values than to high adsorption values, which leads to changing in the error distribution [37].

The energy of adsorption can be described using the Temkin isotherm (Eq. 5). However, this isotherm is valid only for an intermediate range of adsorbate concentrations [38]:

$$q_e = \frac{RT}{b} \ln (A_T C_e) \tag{5}$$

Rearranging Eq. (4) results in Eq. (6):

$$q_e = \frac{RT}{b} \ln (A_T) + \frac{RT}{b} ln (C_e) \tag{6}$$

Plotting q_e versus $\ln(C_e)$ gives a linear regression where the slope for that plot gives the Temkin isotherm constant *(b)* and the intercept gives the Temkin isotherm equilibrium binding constant (A_T) (L/g), where R is the universal gas constant (8.314 J/mol K), T is the temperature in Kelvin (K), and B in Eq. (7) is a constant related to heat of adsorption (J/mol):

$$B = \frac{RT}{b} \tag{7}$$

The Sips isotherm model for mono-component system is a combination between the Freundlich and Langmuir isotherm models. Eq. (8) expresses the Sips model:

$$q_e = \frac{q_{max} (b\,C_e)^{ns}}{(1 + (b\,C_e)^{ns})} \tag{8}$$

where

q_e: The adsorption capacity at equilibrium (mg of adsorbate/g of adsorbent).

C_e: The residual adsorbate concentration at equilibrium in solution (mg/L).

q_{max}: The maximum adsorption capacity corresponding to monolayer coverage (mg of analyte adsorbed/g of adsorbent).

b: The Langmuir constant correlated to the adsorption energy (1/mg adsorbate).

ns: The Sips constant for the heterogeneity of binding surface.

As an extension for the Langmuir isotherm, a model with three parameters was established expressed in Eq. (9). That is Redlich-Peterson isotherm:

$$q_e = \frac{a_{RP} (C_e)}{(1 + (b_{RP} C_e^{n_{RP}}))} \tag{9}$$

where C_e (mg/L) is the residual adsorbate concentration at equilibrium in the solution and q_e (mg/g) is the adsorption capacity at equilibrium. However, a_{RP} (1/g) and b_{RP} $(1/mg)^{nRP}$ do not have physical or chemical meaning. The third parameter n_{RP} is dimensionless that gives an idea about the heterogeneity of adsorption sites on the surface of adsorbents [39].

2.2 Kinetic models

Studying the uptake rate of heavy metals is achieved by the adsorption kinetics where the metal ion uptake rate clearly controls residence time of these compounds at the solid-liquid interface, so and in sequence the mechanism of heavy metal adsorption on the biomass materials will be evaluated using the most common kinetic models.

The simplest one which expresses on the proportionality between the metal adsorption and the number of vacant adsorption sites on the surface of adsorbents is Lagergren model (pseudo-first-order). The nonlinear and linear forms of the model are represented in Eqs. (10) and (11), respectively [40]:

$$q_t = q_e\left(1 + e^{-k_1 t}\right) \tag{10}$$

$$\ln(q_e - q_t) = \ln(q_e) - k_1 t \tag{11}$$

where q_t and q_e (mg/g), respectively, are the adsorption capacity at any time (t) and at equilibrium. k_1 (1/min) is the pseudo-first-order rate constant.

The kinetic model that has the correlation between the adsorption of metal ions and the square of active vacant adsorption sites on the surface of adsorbents is called pseudo-second-order rate model (Eq. 12) [38]:

$$q_e = \frac{q_e^2(k_2 t)}{(1 + (k_2 q_e t))} \tag{12}$$

Eq. (8) can be rearranged to be in the following linear form (Eq. 13):

$$\frac{t}{q_t} = \frac{1}{q_e^2 k_2} + \frac{t}{q_e} \tag{13}$$

where q_t and q_e (mg/g), respectively, are the adsorption capacity at any time (t) and at equilibrium. k_2 (g/mg min) is the pseudo-second-order rate constant.

By plotting $\ln(q_e - q_t)$ versus t and t/q_t versus t in the previous equations (Eqs. (11) and (13)), all the adsorption kinetic parameters can be determined from the slope and the intercept.

The influence of mass transfer resistance on binding metal ions on adsorbents was tested using the intra-particle diffusion model (Weber and Morris model) represented in Eq. (14) [41]:

$$q_t = \mathrm{k_{id}}\,\mathrm{t}^{0.5} + \mathrm{C} \tag{14}$$

where q_t (mg/g) is the adsorption capacity at any time (t), k_{id} (mg/g min$^{0.5}$) is the intra-particle diffusion rate constant, and C (mg/g) is a constant related to the thickness of the boundary layer. From plotting of q_t versus the square root of t, the diffusion constant $\mathrm{k_{id}}$ can be calculated. If this plot passes through the origin, then intra-particle diffusion is the only rate-controlling step.

3. Conclusion

Removal of heavy metals from wastewater would provide an exceptional alternative water resource. Algae biomass adsorbents, which utilized for adsorptive removal of heavy metal pollutants from wastewater, show a promising alternative. Different empirical isotherm models for single analyte have been discussed (i.e., Freundlich, Langmuir, Temkin, Sips, and Redlich-Peterson). In a large number of studies, the Freundlich and Langmuir models are the most commonly and widely used isotherm models. The two kinetic models, which are still in a wide use for studying the rate uptake of heavy metals and their bioadsorption from aqueous solutions, are pseudo-first- and pseudo-second-order kinetic models. In chemisorption process, the pseudo-second-order kinetic model is superior to pseudo-first-order model as it takes into account the interaction of adsorbent-adsorbate through their valency forces.

Acknowledgements

The support of the Center for Environment and Water in the research institute of King Fahd University of Petroleum and Minerals King Fahd University of Petroleum and Minerals is highly acknowledged.

Conflict of interest

The author declares that there are no conflicts of interest.

Author details

Mazen K. Nazal
Center for Environment and Water (CEW), Research Institute (RI) at King Fahd University of Petroleum and Minerals (KFUPM), Dhahran, Saudi Arabia

*Address all correspondence to: mazennazal@kfupm.edu.sa

References

[1] Guidelines for Water Reuse U.S. Environmental Protection Agency Office of Wastewater Management. EPA/600/R-12/618|September 2012

[2] U. S. Environmental Protection Agency. Accessed August 2016. http://www.epa.gov/region09/water/recycling/index.html

[3] Web of knowledge at Web of Science website http://apps.webofknowledge.com/ last time checked July 2016

[4] Ozer A, Akkayaa G, Turabik M. Biosorption of acid red 274 (AR 274) on *Enteromorpha prolifera* in a batch system. Journal of Hazardous Materials. 2005;**126**:119-127

[5] Altenora S, Ncibia MC, Emmanuelb E, Gasparda S. Textural characteristics, physiochemical properties and adsorption efficiencies of Caribbean alga *Turbinaria turbinata* and its derived carbonaceous materials for water treatment application. Biochemical Engineering Journal. 2012;**67**:35-44

[6] Vijayaraghavan K, Yun YS. Bacterial biosorbents and biosorption. Biotechnology Advances. 2008;**26**:266-291

[7] Podgorskii VS, Kasatkina TP, Lozovaia OG. Yeasts—Biosorbents of heavy metals. Mikrobiolohichnyĭ Zhurnal. 2004;**66**:91-103

[8] Jeba Sweetly D. Macroalgae as a potentially low-cost biosorbent for heavy metal removal: A review. International Journal of Pharmaceutical and Biological Archives. 2014;**5**(2):17-26

[9] Sheng PX, Tan LH, Chen JP, Ting YP. Biosorption performance of two brown marine algae for removal of chromium and cadmium. Journal of Dispersion Science and Technology. 2008;**25**:681-688

[10] Oliveira RC, Jouannin C, Guibal E, Garcia O. Samarium(III) and praseodymium(III) biosorption on Sargassum sp.: Batch study. Process Biochemistry. March 2011;**46**(3):736-744. ISSN 13595113

[11] Bishnoi NR, Kumar R, Kumar S, Rani S. Biosorption of Cr(III) from aqueous solution using algal biomass spirogyra spp. Journal of Hazardous Materials. 2007;**145**:142-147

[12] Ofer R, Yerachmiel A, Yannai S. Marine macroalgae as biosorbents for cadmium and nickel in water. Water Environmental Research. 2003;**75**:246-253

[13] Jalali R, Ghafourian H, Asef Y, Davarpanah SJ, Sepehr S. Removal and recovery of lead using nonliving biomass of marine algae. Journal of Hazardous Materials. 2002;**92**:253-262

[14] Schiewer S, Wong MH. Ionic strength effects in biosorption of metals by marine algae. Chemosphere. 2000;**41**:271-282

[15] Yang L, Chen JP. Biosorption of hexavalent chromium onto raw and chemically modified Sargassum sp. Bioresource Technology. 2008;**99**(2):297-307. ISSN 0980-8524

[16] Sakamoto N, Kano N, Imaizumi H. Biosorption of uranium and rare earth elements using biomass of algae. Bioinorganic Chemistry and Applications. 2008;**2008**:1-8. ISSN 1565-3633

[17] Vijayaraghavan K, Jegan J, Palanivelu K, Velan M. Biosorption of cobalt (II) and nickel (II) by seaweeds: Batch and column studies. Separation and Purification Technology. 2005;**44**(1):53-59. ISSN 1383-5866

[18] Oliveira RC, Garcia O Jr. Study of biosorption of rare earth metals (La, Nd, Eu, Gd) by Sargassum sp. biomass in batch systems: Physicochemical evaluation of kinetics and adsorption models. Advanced Materials Research. 2009;**71-73**:605-608. ISSN 1022-6680

[19] Freitas OMM, Martins RJE, DelerueMatos CM, Boaventura RAR. Removal of Cd(II), Zn(II) and Pb(II) from aqueous solutions by brown marine macro algae: Kinetic modeling. Journal of Hazardous Materials. 2008;**153**(1-2):493-501. ISSN 0304-3894

[20] Murphy V, Hughes H, McLoughlin P. Cu(II) binding by dried biomass of red, green and brown macroalgae. Water Research. 2007;**41**:731-740

[21] Herrero R, Cordero B, Lodeiro P, ReyCastro C, Vicente MESD. Interactions of cadmium(II) and protons with dead biomass of marine algae Fucus sp. Marine Chemistry. 2006;**99**:106-116

[22] Ghimire KN, Inoue K, Ohto K, Hayashida T. Adsorption study of metal ions onto crosslinked seaweed *Laminaria japonica*. Bioresource Technology. 2008;**99**(1):32-37. ISSN 0980-8524

[23] Luo F, Liu Y, Li X, Xuan Z, Ma J. Biosorption of lead ion by chemically modified biomass of marine brown algae Laminaria japonica. Chemosphere. 2006;**64**:1122-1127

[24] Chaisuksant Y. Biosorption of cadmium (II) and copper (II) by pretreated biomass of marine alga *Gracilaria fisheri*. Environmental Technology. 2003;**24**:1501-1508

[25] Sheng PX, Ting YP, Chen JP, Hong L. Sorption of lead, copper, cadmium, zinc, and nickel by marine algal biomass: Characterization of biosorptive capacity and investigation of mechanisms. Journal of Colloid and Interface Science. 2004;**275**(1):131-141. ISSN 00219797

[26] Carrilho EN, Gilbert TR. Assessing metal sorption on the marine alga *Pilayella littoralis*. Journal of Environmental Monitoring. 2000;**2**:410-415

[27] Gin KY, Tang YZ, Aziz MA. Derivation and application of a new model for heavy metal biosorption by algae. Water Research. 2002;**36**:1313-1323

[28] Deng L, Su Y, Su H, Wang X, Zhu X. Biosorption of copper(II) and lead(II) from aqueous solutions by nonliving green algae *Cladophora fascicularis*: Equilibrium, kinetics and environmental effects. Adsorption. 2006;**12**:267-277

[29] Yun YS, Parck D, Park JM, Volesky B. Biosorption of trivalent chromium on the brown seaweed biomass. Environmental Science and Technology. 2001;**35**:4353-4358

[30] de Souza FB, de Lima Brandão H, Hackbarth FV, de Souza AAU, Boaventura RAR, de Souza SMAGU, et al. Marine macro-alga *Sargassum cymosum* as electron donor for hexavalent chromium reduction to trivalent state in aqueous solutions. Chemical Engineering Journal. 2016;**283**:903-910

[31] Senthilkumar R, Vijayaraghavan K, Thilakavathi M, Iyer PVR, Velan M. Application of seaweeds for the removal of lead from aqueous solution. Biochemical Engineering Journal. 2007;**33**:211-216

[32] Hamdy AA. Biosorption of heavy metals by marine algae. Current Microbiology. 2000;**41**:232-238

[33] Senthilkumar R, Vijayaraghavan K, Thilakavathi M, Iyer PVR, Velan M. Seaweeds for the remediation of

wastewaters contaminated with zinc(II) ions. Journal of Hazardous Materials. 2006;**136**:791-799

[34] Cechinel MAP, Mayer DA, Pozdniakova TA, Mazur LP, Boaventura RAR, de Souza AAU, et al. Removal of metal ions from a petrochemical wastewater using brown macro-algae as natural cation-exchangers. Chemical Engineering Journal. 2016;**286**:1-15

[35] Kinniburgh DG. 1985. ISOTHERM. A Computer Program for Analyzing Adsorption Data. Report WD/ST/85/02. Version 2.2. British Geological Survey, Wallingford. England

[36] Webber TN, Chakravarti RK. Pore and solid diffusion models for fixed bed adsorbers. AIChE Journal. 1974;**20**:228-238

[37] Kinniburgh DG. General purpose adsorption isotherms. Environmental Science & Technology. 1986;**20**:895-904

[38] Tempkin MI, Pyzhev V. Kinetics of ammonia synthesis on promoted iron catalyst, Acta Physico-Chimica. USSR 12. 1940;327-356

[39] Lagergren S. About the theory of so-called adsorption of soluble substances. Kungliga Svenska Vetenskapsakademiens Handlingar. 1898;**24**:1-39

[40] Ho Y, McKay G. A comparison of chemisorption kinetic models applied to pollutant removal on various sorbents. Process Safety and Environmental Protection. 1998;**76**:332-340

[41] Weber WJ, Morris JC. American Society of Civil Engineers. Kinetic of adsorption on carbon from solutions. Journal of the Sanitary Engineering Division. Proceedings of the American Society of Civil Engineers. 1963;**89**(2):31-60

Chapter 9

Adsorption of Heavy Metals on Layered Double Hydroxides (LDHs) Intercalated with Chelating Agents

Naoki Kano and Shuang Zhang

Abstract

Layered double hydroxides (LDHs) are lamellar ionic compounds containing a positively charged layer and exchangeable anions in the interlayer. In this study, LDHs intercalated with chelating agents were synthesized by anion exchange reaction. The materials synthesized in this work were characterized by chemical analysis, Fourier transform infrared spectroscopy (FT-IR), scanning electron microscopy (SEM), and powder X-ray diffraction (XRD) to confirm their properties. Adsorption experiments from aqueous solutions containing known amounts of some heavy metallic ions onto the adsorbent were explored in a batch system. The amount of metallic ions adsorbed by LDHs intercalated with EDTA and precursor LDHs were determined by inductively coupled plasma-atomic emission spectrometry (ICP-AES) and atomic absorption spectrometry (AAS). In order to examine the adsorption capacity of LDHs intercalated with chelating agents, the adsorption experiment was investigated under the optimum condition. The data were applied to Langmuir and Freundlich isotherm models. The pseudo-second-order kinetic model was more adequate to describe the kinetic in this case. LDHs intercalated with chelating agents synthesized in this work can be promising adsorbents for heavy metals. It is very significant information from the viewpoint of environmental protection.

Keywords: adsorption, layered double hydroxides (LDHs), chelating agent, heavy metals, adsorption isotherms, kinetics

1. Introduction

Recently, environmental pollution is one of the most serious problems in the world due to its deep effect on the future of human beings. Then the investigation for resolving the problem of the environmental pollution in the world began to draw major public attention [1, 2]. It is well known that the major forms of environmental pollution include air pollution, water pollution, soil pollution, and so on. Among them, water pollution is the most serious due to its liquidity which may bring other pollutions. Surface water pollution and groundwater contamination are some of the environmental problems today. One of the cases of environmental pollution is due to heavy metal contaminants such as copper, lead, cadmium, chromium, arsenic, zinc, etc. Also, heavy metals are concerned because of their strong toxicity even at low concentrations. Based on the type of mining, the kinds and the concentrations of metal ions are many and varied. Heavy metal ions have high toxicity and poor biodegradability for plants and animals at higher concentrations [3, 4].

In recent years, clay minerals have been aroused increasing interest as adsorbents by virtue of their properties, which make them attractive materials for adsorbing heavy metal ions. Their abundance in nature, low cost, and good cation adsorptive properties, a result of their negatively charged layers and high specific surface areas, make them suitable for adsorption of metal ions [5, 6]. LDH used in this paper are the antitypes of clay minerals. Layered double hydroxides (LDHs) are lamellar ionic compounds containing a positively charged layer and exchangeable anions in the interlayer. They consist of brucite-like layers and are represented by the general formula $[M^{II}_{1-x}M^{III}_{x}(OH)_2](A^{n-})_{x/n}\cdot mH_2O$, where cationic M^{II} and M^{III} are divalent and trivalent metals and occupy the octahedral holes in the brucite-like layer. An^{-} is the interlayer exchangeable anions, which is located in the hydrate layered galleries, and x is the layer charge density $x = [M^{II}]/([M^{III}] + [M^{III}])$ [7–9]. The layered structure of LDHs is shown in **Figure 1**.

The ethylenediaminetetraacetic acid (EDTA) is a chelating agent widely used in industry and agriculture. It forms strong complexes with the ratio 1:1 between heavy metal ions and ligand. The EDDS (N, N′-1, 2-Ethanediylbis-1-Aspartic Acid) is also a chelating agent, which may offer a biodegradable alternative to EDTA and is currently used on a large scale in numerous applications [10, 11]. The structure of this two chelating agents were shown in **Figure 2**.

Considering the structure of LDHs, it is suggested that these compounds can be intercalated with different polydentate ligands. Recently, the study using LDHs modified with chelating agents as the potential adsorbents of heavy metals from aqueous solution has been reported [12, 13]. The aim of this work is at first to synthesize and to characterize LDHs intercalated with EDTA or EDDS and to study the uptake of heavy metals (Cu^{2+}, Pb^{2+}, Cd^{2+}) by these hybrid compounds. The following five kinds of compounds synthesized in this work are ZnAl-NO_3 (L1), ZnAl-EDTA (L2), MgAl-NO_3 (L3), MgAl-EDTA (L4), and MgAl-EDDS (L5). To confirm the effect of intercalation with EDTA, the adsorption of metallic ions onto L1 and L2 is also compared. This study investigated the adsorption ability of LDHS as adsorbent for Pb, Cu, and Cd from aqueous solution. Finally, the further

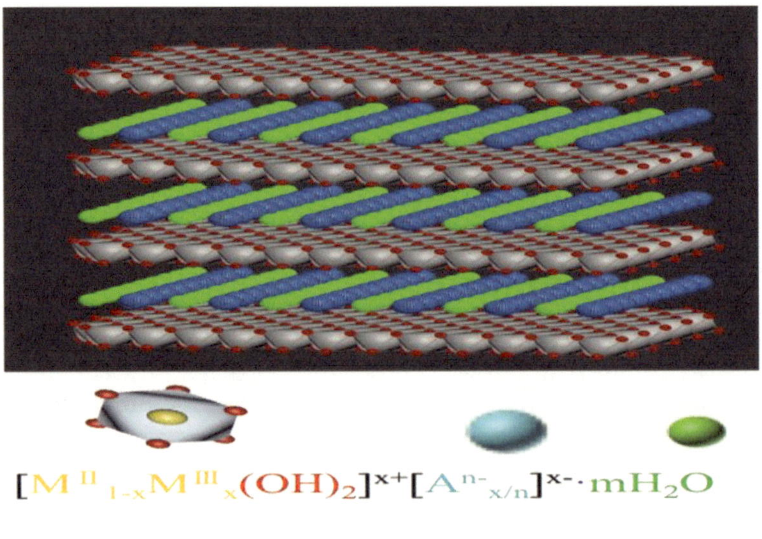

Figure 1.
The layered structure of LDHs.

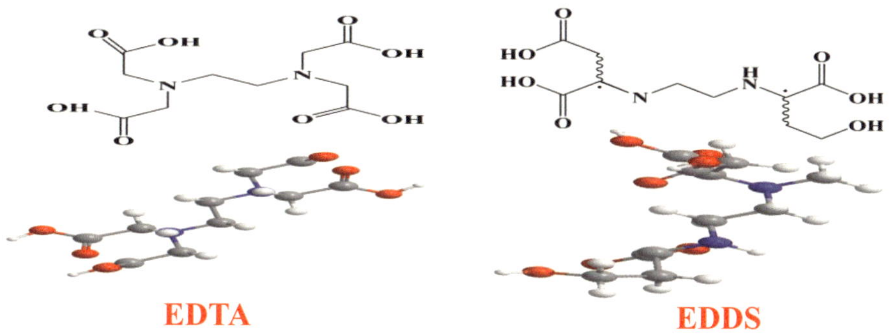

Figure 2.
The structure of EDTA and EDDS.

developments of LDHs as useful adsorbent with the future of application in the environmental chemistry are mentioned.

2. Experimental section

2.1 Materials and reagents

Chemical reagents including $Zn(NO_3)_2{\cdot}6H_2O$, $Al(NO_3)_3{\cdot}9H_2O$, $Mg(NO_3)_2{\cdot}6H_2O$, $Cu(NO_3)_2{\cdot}6H_2O$, $Pb(NO_3)_2$, $Na_2H_2EDTA{\cdot}2H_2O$, NaOH, HNO_3 and Zn(II) Mg(II), and Al(II) standard solution were purchased from Kanto Chemical Co., Inc.; Cd(II) standard solutions were prepared by diluting a standard solution (1000 mg L^{-1}); EDDS (35%) was purchased from Sigma Co., Ltd.; and all reagents used were of analytical grade. CO_2 free water (>18.2 MΩ) which was treated as an ultrapure water system (RFU 424TA, Advantech Aquarius) was employed throughout the work. The pH meter (HORIBA F-72) was used for measurement of pH while adjusting the pH by using 0.01 or 0.1 mol L^{-1} NaOH aqueous solution and 0.01 or 0.1 mol L^{-1} HNO_3 aqueous solution. All synthesis should be performed under a N_2 atmosphere condition to avoid carbonate contamination.

2.2 Synthesis of the adsorbents

The synthesis of LDHs intercalated with EDTA or EDDS includes two steps: (1) the preparation of the precursor LDHs (L1 or L4) and (2) the anion exchange reaction of this compound with chelating agents [14]. All the synthesis was purged with N_2 to avoid CO_2 uptake from atmosphere.

- Synthesis of Precursor L1 and L4

L1 was prepared by dropping addition of 100 mL aqueous solution of 0.02 mol L^{-1} $Zn(NO_3)_2{\cdot}6H_2O$ and 0.01 mol L^{-1}Al $(NO_3)_3{\cdot}9H_2O$ to 100 mL NaOH/$NaNO_3$ solution. Then, the solutions were agitated at 70°C for 8 h by maintaining the pH, separated by centrifugation, and washed until neutral. L4 was also synthesized by using Mg $(NO_3)_2{\cdot}6H_2O$ and Al $(NO_3)_3{\cdot}9H_2O$ as the similar method [15, 16].

- Synthesis of L2, L3, and L5

L2 was synthesized as follows. Under a N_2 atmosphere, 0.015 mol of EDTA or EDDS was added to the 150 mL of suspended solution of L1. Then, the mixing solutions were agitated at 70°C for 8 h under a certain pH degree, then separated by centrifugation, washed until neutral, and then dried at 60°C overnight [12, 17]. L5 was synthesized by L4 as the similar method for L2.

2.3 Characterization of these adsorbents

Elemental chemical analyses of C, H, and N in LDHs were carried out using an elemental analyzer instrument (JMC10, J-SCIENCE LAB CO., Ltd.). After dissolving the sample by HNO_3, the amount of metallic ions in LDHs was obtained by ICP-MS (Agilent HP 4500, Thermo). Infrared spectra were obtained using the KBr disc method, with wavenumbers from 400 to 4000 cm^{-1} on a FT-IR (FTIR-4200, Jasco, Japan). XRD (X-ray powder diffraction) of LDHs samples were carried out on a RINT2500HR-PC (RIGAKU Corporation) using Cu $K\alpha$ radiation in the scanning range of 2–80°. N_2 adsorption and desorption isotherms were employed to determine the specific surface area by the specific surface area analyzers (AUTOSORB-1, Quabtachrome Inc., USA). The surface morphology of LDHs was surveyed using scanning electron microscopy (SEM; JSM-5800, JEOL, Japan). The element distribution and the component analysis were also analyzed by electron probe micro analyzer (EPMA; 1600, Shimadzu Corporation).

2.4 Adsorption experiments

For obtaining the optimum conditions regarding the adsorption of heavy metal, the batch experiments were studied by varying pH, contact time, adsorbent dose, and initial concentration on the adsorption of heavy metal [18–20]. The adsorption experiments of Cu(II) and Pb(II) using L2 and L3 were carried out. A certain amount of L2 or L3 was contacted with 30 mL of an aqueous solution containing known initial each metal ion (nitrate salts) ranging from 0.1 to 2 g L^{-1}. Sorption experiments were conducted in the pH range of 2–6, contact time from 30 minutes to 6 h, temperature from 25 to 40°C, and adsorbent dosage 5–40 mg. The pH of each solution was adjusted using 0.1 mol L^{-1} NH_4OH and 0.1 mol L^{-1} HNO_3. The adsorption capacities of Cu(II) or Pb(II) on L1, L2, and L3 were compared with that of commercial LDHs: DHT-4A ([$Mg_{4.5}Al_2(OH)_{13}CO_3 \cdot 3.5H_2O$], Kyowa Chemical Industry Co., Ltd), which is abbreviated as L0 below.

The adsorption experiments of Cd(II) were also carried out similar as the method below. The experiment using heavy metallic ions solution without the adsorbent was also performed to identify potential loss of heavy metallic ions during the process such as precipitation. To confirm the effect of intercalation with EDTA, the adsorption of Cu(II), Pb(II) Cd(II) onto L4 and L5 are also compared.

The suspension containing the adsorbent and each of the above metallic solution was filtered through a 0.10 μm membrane filter (Mixed Cellulose Ester 47 mm, Advantec MFS, Inc.) to remove each metallic ion that have been adsorbed into the adsorbent. Then, the concentration of Cu(II) or Pb(II) in the filtrate was determined with an atomic absorption spectrophotometer (AAS), and the concentration of Cd(II) in the filtrate was determined by inductively coupled plasma-atomic emission spectrophotometer (ICP-AES) (SPS 1500, Seiko Instrument Inc).

2.5 Data analysis

For data analysis, various equilibrium, kinetic, and thermodynamic models (equations) were employed to interpret the data and establish the extent of adsorption. The metallic ions uptake by each adsorbent was calculated using the Eq. (1):

$$Q = \frac{(C_0 - C_e)}{W} \cdot V \quad [\mu g \cdot g^{-1}] \tag{1}$$

where Q is the adsorption capacities at equilibrium (μg g^{-1}), c_o and c_e are the initial and equilibrium concentrations of metallic ions in a batch system, respectively (mg L^{-1}), V is the volume of the solution (L), and W is the weight of each adsorbent (g) [15, 16].

2.6 Adsorption isotherms

In adsorption processes, it is necessary and critical for the equilibrium isotherm studies to predict the behavior of pollutant adsorption onto the sorbent surfaces. Two common adsorption models, Langmuir and Freundlich isotherm models, were applied to evaluate the adsorption data obtained in this study.

The Langmuir adsorption model is based on the assumption that maximum adsorption corresponds to saturated monolayer of solute molecules on the adsorbent surface [21, 22]. Langmuir model is given by Eq. (2):

$$\frac{C_e}{q_e} = \frac{C_e}{q_{\max}} + \frac{1}{K_L q_{\max}} \tag{2}$$

where C_e is the concentration of metallic ions in batch system at equilibrium (mg L^{-1}), q_e is the amount of adsorption of metallic ions at equilibrium (mg g^{-1}), q_{max} is the maximum adsorption capacity on the surface of adsorbent (mg g^{-1}), and K_L is the equilibrium adsorption constant (L mg^{-1}). A plot of C_e/q_e versus C_e gives a straight line ($Y = A + BX$) with slope of $1/q_{max}$, and the intercept is $1/(K_L\, q_{max})$;

The linearized Freundlich model isotherm is represented by the following equation:

$$\lg q_e = \lg K_F + (1/n) \lg C_e \tag{3}$$

where K_F and $1/n$ indicate the adsorption capacity and the adsorption intensity of the system, respectively. The plots of q_e versus C_e in log scale can be plotted to determine values of $1/n$ and K_F depicting the constants of Freundlich model. The greater the value of the n, the more favorable is the adsorption [15, 16, 23].

2.7 Kinetic model

The kinetic data can be used to determine the time required for adsorption equilibrium and provide useful data to improve the efficiency of the adsorption model and develop predictive models [24, 25]. In this work, pseudo-first-order and pseudo-second-order models were applied for modeling the adsorption process. The pseudo-first-order model is expressed as the Eq. (4):

$$\ln(q_e - q_t) = \ln q_e - k_1 t \tag{4}$$

where q_t and q_e (μg g^{-1}) are the metal amount adsorbed at time at t (h) and equilibrium, respectively, and k_1 is the rate constant of the pseudo-first-order adsorption (h^{-1}).

The linear form of the pseudo-second-order rate equation is given as follows:

$$\frac{t}{q_t} = \frac{1}{k_2 q_e^{\,2}} + \frac{t}{q_e} \tag{5}$$

where k_2 (g $μg^{-1}$ h^{-1}) is the pseudo-second-order rate constant of the adsorption [15, 16, 25].

3. Results and discussion

3.1 Chemical analysis

The chemical analysis of LDH samples is shown in **Table 1**. The molar ratio of M^{II}/M^{III} in L1 or L4 is nearly 2 which is well fitted to the expected formula. However, L2, L3, or L5 has lower M^{II}/M^{III} ratio than L1 or L4. The presence of polydentate ligand (Zn-EDTA or Al-EDTA complex) can result in this decrease of the ratio. Moreover, the decrease suggests that octahedron in hydroxyl layer has a partial dissolution (pK_{sp} ($Zn(OH)_2$) = 13.7, pK_{sp} ($Al(OH)_3$) = 32.7, pK_{sp} ($Mg(OH)_2$) = 12.7) during the anion exchange reaction which is performed at pH 5–6 [12, 17, 26]. The C/N of L5 is lower than that of EDTA ligand (5), and the little gap between them may be mainly due to the registration of nitrate ions in the interlayer. The more nitrate ions are included in LDHs, the lower the C/N value is [27].

3.2 FT-IR spectra

The FT-IR spectra of L1 and L2 and L3 are shown in **Figure 3**, and that of L4 and L5 are shown in **Figure 4**. Typical M-OH (M—metallic ions) vibration modes due to the hydroxide layer between 400 and 1000 cm^{-1} are found in both **Figures 3** and **4**.

The very sharp peak at 1385 cm^{-1} in **Figures 3(a)** and **4(a)** is attributed to the NO_3^- stretching vibration. The NO_3^- stretching vibration at 1385 cm^{-1} is not observed from **Figures 3(b)** to **4(b)**. It may be due to the group which is hidden by the band at 1394 cm^{-1} [26, 28]. The absorption bands at 1600 and 1394 cm^{-1} are characteristics of the symmetrical and asymmetrical vibration of COO- groups. The position of these bonds is similar to the spectrum of LDHs which is reported by Parida et al. [29] and [30]. It is found that EDTA has been intercalated into the interlayer successfully, although a certain amount of $-NO_3$ may still retain in the compound judging from the results of chemical analysis. The wide band at around 3450 cm^{-1} may be attributed to the -H bonding stretching vibrations of -OH groups and water molecules. The band at 1623 cm^{-1} of L1 and L4 is assigned to water bending vibration [8, 26].

3.3 XRD patterns

XRD patterns of L1 and L2 and L3 are shown in **Figure 5**, and those of L4 and L5 are shown in **Figure 6**. They are typical XRD patterns of LDHs. The strong diffraction peaks at low angle, assigned to basal planes (003), (006), (009), were sharp and symmetric compared to the peaks at high angle, which are characteristics of clay mineral

	wt%	N	H	Atomic ratios			Proposed formula
		C		M^{II}/ M^{II}	C/H	H/N	
L1	0.54	5.17	3.24	2.23	0.01	8.79	$[Mg_2Al(OH)_6]NO_3$
L2	11.2	3.83	4.19	1.88	0.22	20.1	$[Mg_2Al(OH)_6]_2[C_{10}H_{14}N_2O_8]$
L3	9.82	2.21	4.15	1.79	0.20	26.3	$[Mg_2Al(OH)_6]_2[C_{10}H_{13}N_2NaO_8]$
L4	0.06	4.26	2.38	2.10	0.00	7.82	$[Zn_2Al(OH)_6]NO_3$
L5	13.7	3.35	3.60	1.67	0.31	15.1	$[Zn_2Al(OH)_6]_2$ $[C_{10}H_{14}N_2O_8]$

Table 1.
Chemical analysis results of L1, L2, L3, L4, and L5.

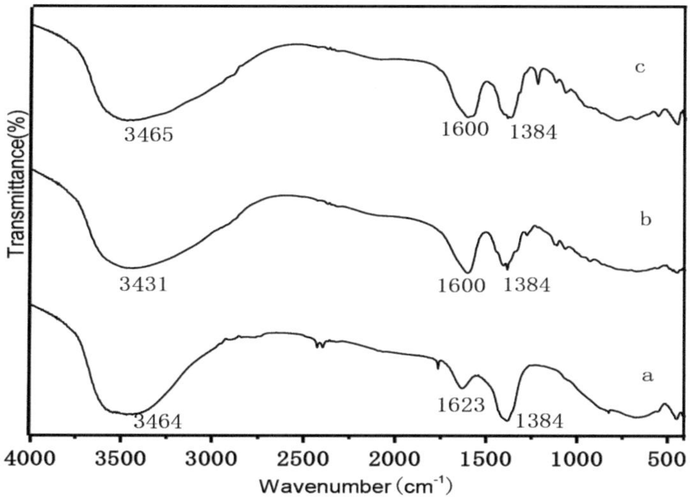

Figure 3.
FT-IR spectra of (a) L1, (b) L2, and (c) L3.

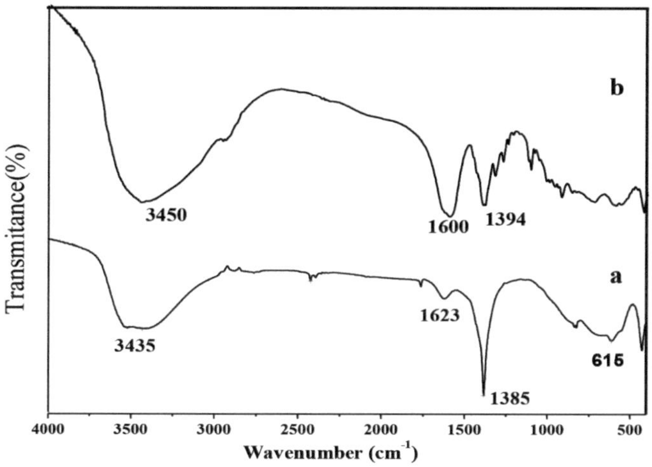

Figure 4.
FT-IR spectra of (a) L4 and (b) L5.

shaving a layered structure [29–31]. From the XRD pattern, the basal spacing (d) values of sample were calculated by using Bragg equation and the angle of peak (003).

Then the gallery height was obtained by subtraction from the basal spacing to the layer width (0.48 nm) [30]. The basal spacing and the gallery height of L1, L2, L4, and L5 are shown in **Table 2**. It indicates that the intercalation of EDTA into NO_3-LDHs gives rise to an increase of basal spacing. This basal spacing could identify the existence of EDTA, because it is close to the dimensions of EDTA complexes (0.9 nm^{-1} nm) founded by single crystal XRD of M-EDTA (M—metallic ions) compound [12, 17, 30, 32].

3.4 SEM micrographs

SEM images of all composite synthesized in this work are shown in **Figure 7**. These adsorbents have clear plate-like morphology, which is typical for LDHs [33]. The intercalated product particles are more homogeneous than the precursor product

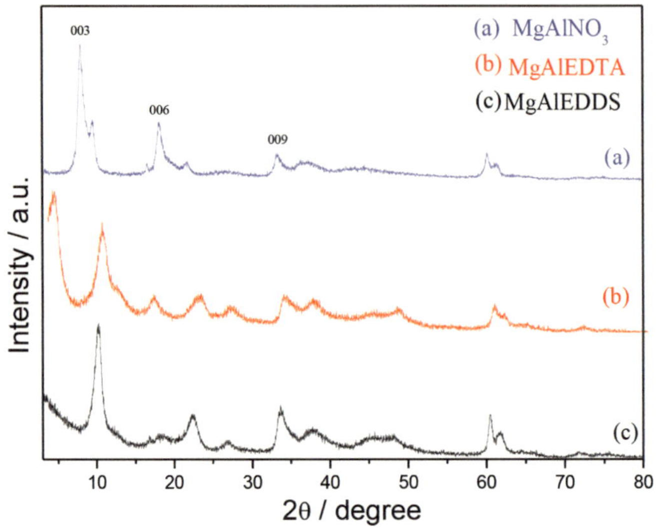

Figure 5.
XRD patterns of (a) L1, (b) L2, and (c) L3.

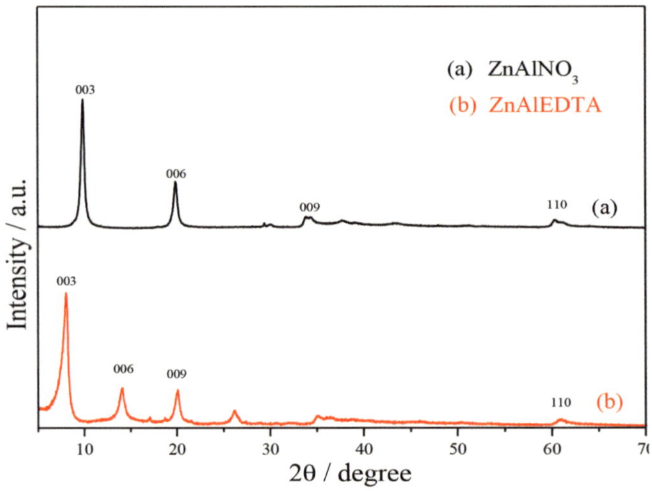

Figure 6.
XRD patterns of (a) L4 and (b) L5.

	L1	L2	L4	L5
Basal spacing	0.91	1.42	0.89	1.47
Gallery height	0.43	0.94	0.41	0.99

Table 2.
The basal spacing of L1, L2, L4, and L5 calculated from XRD by using Bragg's equation.

which may be due to the hydrogen bonding on the layer. Hydrogen bonding makes soft agglomeration occur on the surface of the LDHs, and after the chelating agent, anion replaces the nitrate ions between the layers, the hydrogen bonding between the hydroxyl groups is reduced, and the aggregation is weakened to a certain extent.

The inhomogeneous surface of the adsorbent indicated that a large amount of metal salt attached to the surface of the hydrotalcite in an excessive state, its unique

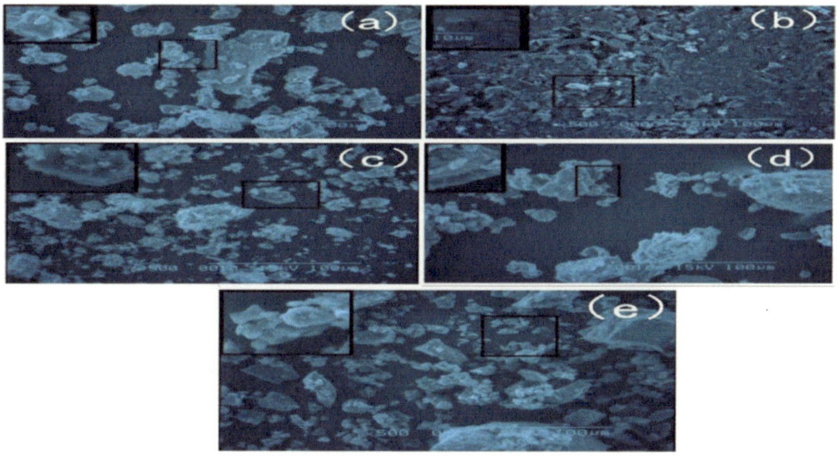

Figure 7.
SEM image of (a) L1, (b) L2, (c) L3, (d) L4, and (e) L5.

layered structure, resulting in removal of heavy metal ions in the aqueous solution not only by interlayer anion and heavy metal cation interaction but also rely on the role of surface adsorption and sedimentation.

3.5 Electron probe microanalyzer (EPMA)

Element distribution analysis of L1, L2, L4, and L5 by EPMA is shown in **Figure 8**. After the ion exchange, the element distribution of N decreased obviously (by comparing red parts in these pictures), and this decrease is observed in both MgAl-LDHs (a, b) and ZnAl-LDHs (c, d). Furthermore, it is found that the moles of divalent metals are at least equal to or greater than that of the trivalent metals [34, 35], which is consistent with the results of chemical analysis.

3.6 Specific surface area

Figure 9 has shown the specific surface area of the product. Specific surface area of L2 and L3 are bigger than that of L1, and that of L5 is bigger than L4, which may be attributed to intercalation of EDTA or EDDS. Specific surface area of L0 is bigger than that of L1; it is due to the difference of their particle size.

3.7 Adsorption of Cu(II) or Pb(II) onto L1, L2, L3, and L0

The adsorption capacities of Cu(II) or Pb(II) onto L1, L2, L3, and L0 are compared in **Figure 10**. The adsorption efficiency of Cu^{2+} was larger than that of Pb^{2+} for the same absorbent, which could be attributed to their stability constant (EDTA-Cu, 18.7; EDDS-Cu, 18.4; EDTA-Pb, 18.0; EDDS-Pb, 12.7) [12, 26]. That is to say, it can be considered that the large adsorption capacity is obtained when the stability constant of chelate-metal is high. By comparing among adsorbents used in this work, the order of the adsorption capacity is L2 > L3 > L0 > L1. The higher adsorption efficiency of L0 than L1 may be attributable to its high specific surface area.

3.8 Adsorption of Cu(II), Pb(II), and Cd(II) onto L4 and L5

In order to confirm the effect of the intercalation with chelate agents on the adsorption capacity of metals, the adsorption experiments of some metallic elements

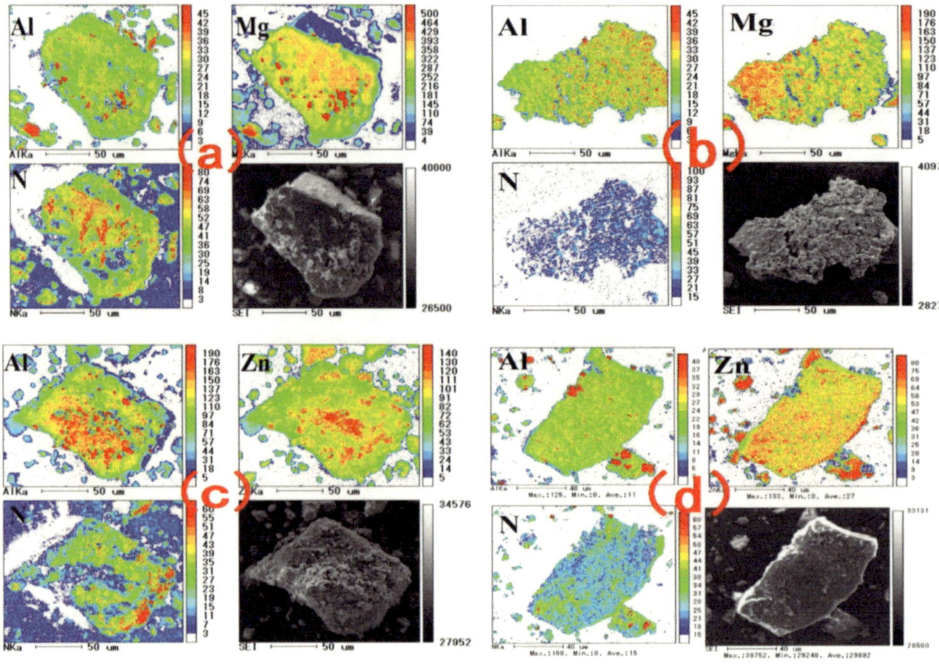

Figure 8.
Element distribution analyzed by EPMA of (a) L1, (b) L2, (c) L4, and (d) L5.

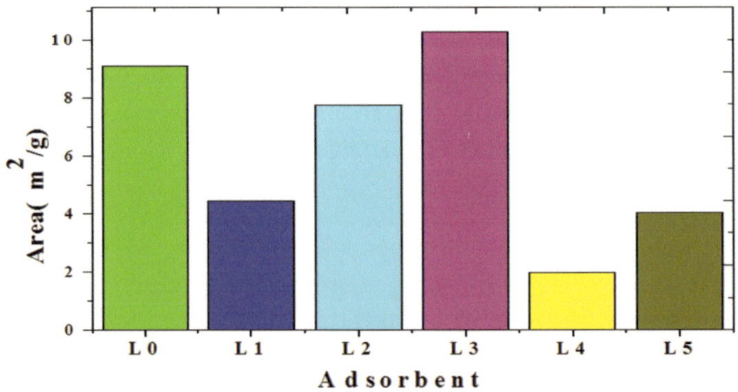

Figure 9.
Specific surface area of L0, L1, L2, L3, L4, and L5 by BET method.

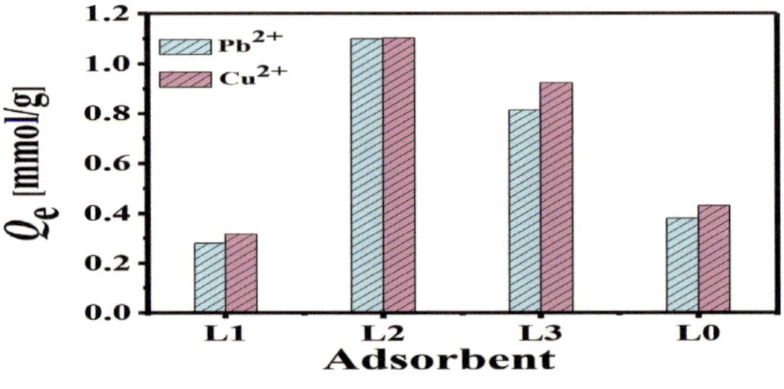

Figure 10.
The adsorption capacity of Pb^{2+} and Cu^{2+} onto L1, L2, L3, and L0.

onto L1 and L2 are compared. The adsorption of Cd(II), Cu(II), and Pb(II) onto these LDHs under the optimum condition are shown in **Figures 11–13**, respectively.

Both LDHs were found to take up Cd(II), Cu(II), and Pb(II) from aqueous solutions, and the uptake was found to increase with time. The adsorption capacity of both LDHs for Cd(II), Cu(II), and Pb(II) increased rapidly during the initial stages, and thereafter it increased gradually. It is generally found that the time needed for L5 to reach equilibrium was shorter than that for L4. From the adsorption experiment, the improvement of adsorption capacity by intercalation was observed. On the other hand, the adsorption capacity of Cu(II) and Pb(II) at equilibrium was

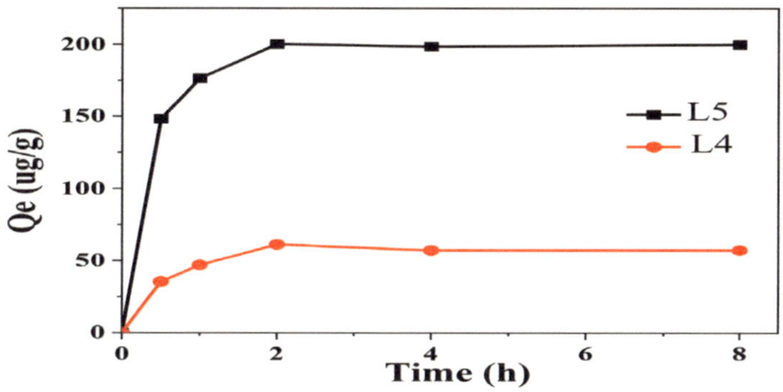

Figure 11.
Adsorption of Cd(II) onto L4 and L5.

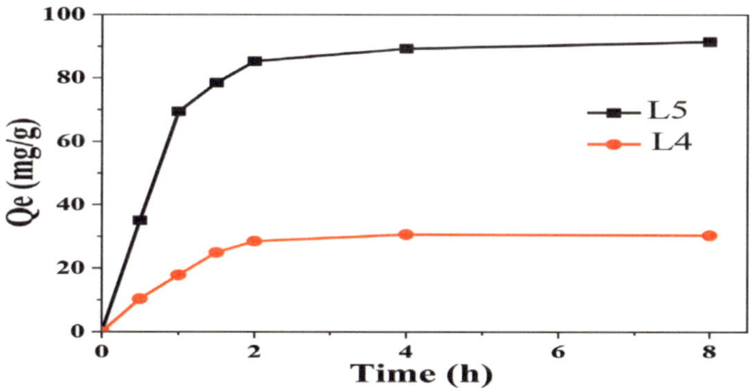

Figure 12.
Adsorption of Cu(II) onto L4 and L5.

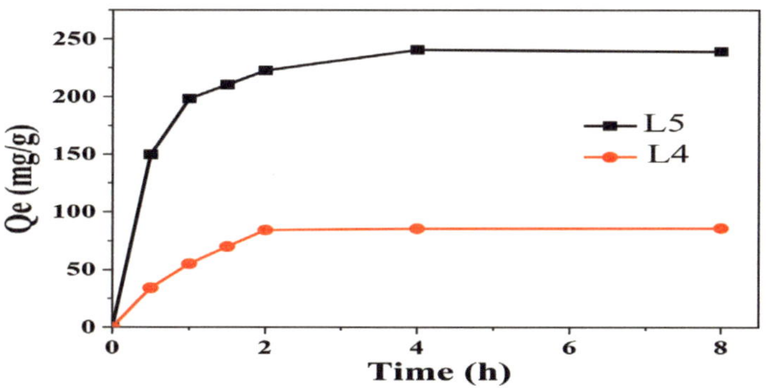

Figure 13.
Adsorption of Pb(II) onto L4 and L5.

higher than that of Cd(II). It is considered that heavy metal was removed by LDHs including two mechanisms: chemical precipitation and chelation [16]. In the first case, the hydroxyl anions compete with chelating agents for the precipitation of metal hydroxides at higher pH, and divalent ions are usually selectively dissolved. In the second case, the adsorption affinity is generally determined by the stability constant of the corresponding complex [36–38].

3.9 Adsorption isotherms

The adsorption isotherms for Cu(II) or Pb(II) were obtained under the optimum adsorption conditions (i.e., pH 6, contact time 120 minutes, temperature 25°C, and adsorbent dosage 10 mg). The adsorption isotherms of Cu(II) or Pb(II) onto L2 and L3 were analyzed using Langmuir and Freundlich equations and were shown in **Figures 14** and **15**, respectively. From **Figure 14**, the linear correlation coefficient (R^2) of L2 (Pb^{2+}), L3 (Pb^{2+}), L2 (Cu^{2+}), and L3 (Cu^{2+}) and other parameter for Langmuir and Freundlich isotherms model were shown in **Table 3**, respectively. That is, L2 and L3 synthesized in this work are well fitted by Freundlich adsorption isotherms models.

3.10 Kinetic model

The kinetic isotherms for Cu(II) or Pb(II) were obtained under the optimum adsorption conditions (i.e., pH 6, concentration 200 ppm, temperature 25°C and

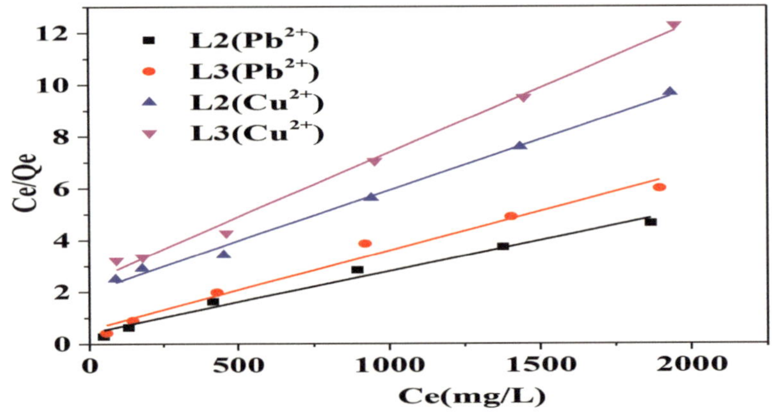

Figure 14.
The correlation of experimental data to Langmuir isotherm models.

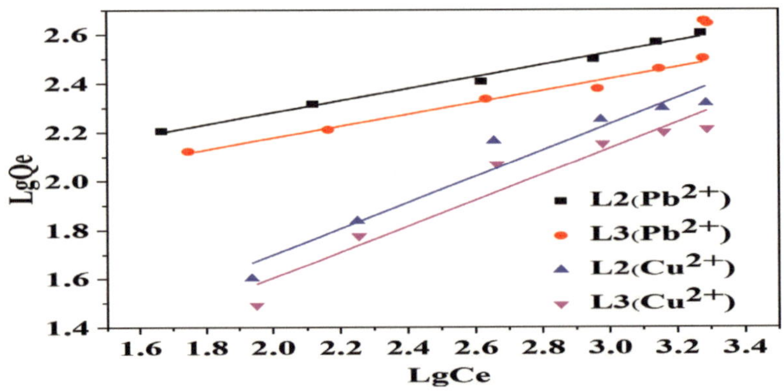

Figure 15.
The correlation of experimental data to Freundlich isotherms models.

Sample/T (298 K)	Langmuir			Freundlich		
	R^2	K_L (L^{-1} mg^{-1})	q_{max} (mg g^{-1})	R^2	K_F ($mg^{1-1/n}$ g^{-1} L^{-1})	n
L2(Pb^{2+})	0.979	5.60×10^{-3}	422	0.988	346	4.11
L3(Pb^{2+})	0.976	5.40×10^{-3}	330	0.983	199	4.17
L2(Cu^{2+})	0.994	1.90×10^{-3}	256	0.930	9.90×10^{-3}	1.87
L3(Cu^{2+})	0.993	2.10×10^{-3}	201	0.914	2.20×10^{-3}	1.89

Table 3.
Coefficient of Langmuir and Freundlich isotherms for Cu(II) and Pb(II) adsorption onto L2 or L3.

adsorbent dosage 10 mg). The parameters for two kinetic models of adsorption of Cu(II) or Pb(II) on L2 or L3 are presented in **Table 4** which showed that adsorption process followed pseudo-second-order rather than pseudo-first-order model.

The second order kinetic models plot for the adsorption of Cu(II) or Pb(II) on L2 or L3 is shown in **Figure 16**. The experimentally calculated values of q_e at various concentrations were in a good agreement with theoretical calculated values. Also, the values of correlation coefficients (R^2) for the pseudo-second-order kinetic model was nearly 1, indicated that pseudo-second-order kinetic model was better obeyed.

3.11 Comparison of the adsorption capacities of the materials with other sorbents

The comparison of maximum adsorption capacity of these LDHs for Cu(II) in a present study with that of another adsorbents in previous literatures [39] are

Sample/T (298 K)	$q_{e\,EXP}$ (mg g^{-1})	Pseudo-first-order			Pseudo-second-order		
		R^2	q_e (mg g^{-1})	K_1 (h^{-1})	R^2	q_e (mg g^{-1})	K_2 (g mg^{-1} h^{-1})
L2(Pb^{2+})	228	0.990	217	5.48	0.997	276	2.75
L3(Pb^{2+})	169	0.940	169	5.27	0.993	229	1.25
L2(Cu^{2+})	71	0.983	78.5	4.35	0.991	111	0.104
L3(Cu^{2+})	59	0.983	54.8	4.09	0.995	91.9	0.910

Table 4.
The kinetic fit parameters for Cu(II) and Pb(II) adsorbed on L2 or L3.

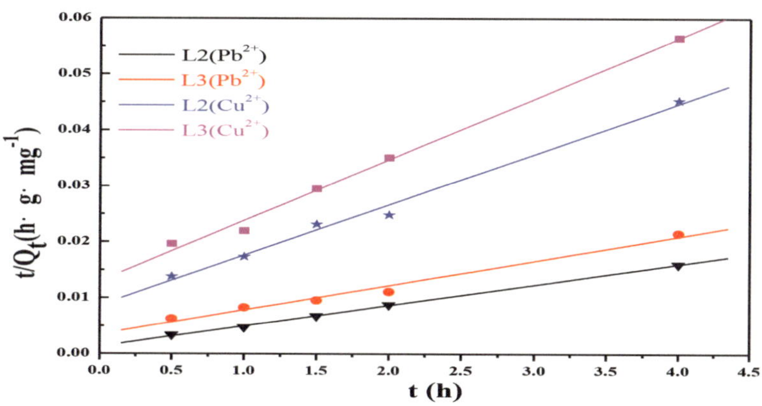

Figure 16.
The correlation of experimental data to pseudo-second-order models.

Adsorbents	Q (mg g^{-1})	References
LS-LDH: MgAl-LDH intercalated by sulfonated lignin (LS)	64	[39]
TA-HTC: hydrotalcite modified by tannin	81	[40]
CL-LDH: LDHs intercalated by chloride	38	[41]
H100-LDH: MgAl-LDH intercalated by humate anions	85	[12]
Sx-LDH: MgAl-LDH intercalated by polysulfide	127	[42]
L0 (MgAl-CO_3)	28	This study
L1 (MgAl-NO_3)	20	This study
L2 (MgAl-EDTA)	71	This study
L3 (MgAl-EDDS)	59	This study
L4 (ZnAl-NO_3)	31	This study
L5 (ZnAl-EDTA)	90	This study

Table 5.
Comparison of the adsorption capacities of LDHs in other literature Cu^{2+}.

Adsorbents	Q (mg g^{-1})	References
MNP-CTS—MNPs modified with chitosan (CTS)	140	[43]
CL-LDH—LDHs intercalated by chloride	40	[16]
CDpoly-MNPs—(CM-β-CD) polymer-modified Fe_3O_4 nanoparticles	65	[44]
H100-LDH—MgAl-LDH intercalated by humate anions	99	[12]
MoS_4-LDH—LDHs intercalated with the MoS_4^{2-} ion	290	[45]
L0 (MgAl-CO_3)	78	This study
L1 (MgAl-NO_3)	58	This study
L2 (MgAl-EDTA)	228	This study
L3 (MgAl-EDDS)	169	This study
L4 (ZnAl-NO_3)	80	This study
L5 (ZnAl-EDTA)	223	This study

Table 6.
Comparison of the adsorption capacities of LDHs in other literature Pb^{2+}.

presented in **Table 5**. Moreover, **Table 6** shows the comparison of adsorption capacity of Pb(II) by other adsorbents reported in the literature. As seen in **Tables 5** and **6**, the adsorption capacity of these LDHs for Cu(II) and Pb(II) in this work is on a level with that of another adsorbents in previous works.

4. Conclusions

In present study, LDHs intercalated with chelating agents have been extensively examined and applied for adsorption of aqueous containing heavy metals and REEs. The following five kinds of compounds were synthesized (MgAl-NO_3 (L1), MgAl-EDTA (L2) and MgAl-EDDS (L3), ZnAl-NO_3 (L4), ZnAl-EDTA (L5)). These five kinds of synthesized samples are characterized by some instruments and the adsorption capacities of LDHs intercalated with chelating agents for Cu(III), Pb(III), and Cd(III), and REEs ions were investigated by batch experiments. Influence of various condition including pH, adsorbents dose, concentration of metallic ions, adsorption time, and temperature on the removal of metallic ions was evaluated. The Langmuir and Freundlich models were used for the mathematical description of the adsorption isotherms. The suitability of the kinetic model for the adsorption processes is also discussed. The following matters were suggested from the experimental results:

1. In present study, the precursor LDHs (ZnAl-NO_3 and MgAl-NO_3) were intercalated with the chelating agent EDTA (ethylenediaminetetraacetic acid) and EDDS (N, N′-1, 2-Ethanediylbis-1-Aspartic Acid) by anion exchange. The obtained material was characterized and used for the removal of heavy metallic ions and REEs removal from aqueous solutions. The result from FT-IR etc. suggests that the intercalation into LDHs is performed successfully.

2. LDHs synthesized in this work were very effective for removing heavy metallic ions from water solutions. Higher adsorption efficiency is obtained by intercalating chelating agent (i.e., EDTA or EDDS) into LDHs. It is considered that the adsorption capacity of metallic ions onto LDHs is based on the stability constant of metal-chelating agents. For example, the adsorption efficiency of Cu(III) was higher than that of Pb(III) for the same absorbent.

3. Adsorption isotherms of adsorption data were studied at varying initial concentration of metallic ions under optimized conditions of contact time and the dosage of adsorbents in this work. The adsorption experimental data of heavy metallic ions onto LDHs were well fitted by the Freundlich adsorption isotherms model. The results suggest that LDHs synthesized in this work could be suitable as sorbent materials for the adsorption and removal of heavy metal ions from aqueous solutions.

4. The pseudo-first-order kinetic and pseudo-second-order models were applied to test the experimental data and explain the kinetics of the LDHs adsorption process. The comparison of evaluated correlation coefficients suggested that the pseudo-second-order model is most suitable for describing the adsorption processes. The confirmation of this model implies that the rate-limiting step in this adsorption system may be controlled by chemical process. Also, the concentrations of both adsorbent and adsorbate are associated with the rate determining step of the adsorption process.

From this work, it was quantitatively clarified that LDHs could be an efficient adsorbent for heavy metal. It is a very significant information from the viewpoint of environmental protection and can be used for treating industrial waste waters including pollutants and thus a promising option for the treatment of contaminated waters.

Acknowledgements

The present work was partially supported by a Grant-in-Aid for Scientific Research from the Japan Society for the Promotion of Science (Research Program(C), no. 16K00599) and a fund for the promotion of Niigata University

KAAB Projects from the Ministry of Education, Culture, Sports, Science and Technology, Japan. The authors are also grateful to Mr. M. Ohizumi of the Office for Environment and Safety in Niigata University, Dr. E. Tayama of Faculty of Science, Dr. M. Teraguchi, Mr. T. Nomoto, Prof. T. Tanaka, and Mr. T. Hatamachi of Faculty of Engineering and Mr. M. Kobayashi of Facility of Dentistry in Niigata University for permitting the use of ICP-AES, elemental analyzer, FT-IR, SEM, specific surface area analyzers, and EPMA and for giving helpful advice in measurement.

Author details

Naoki Kano[1*] and Shuang Zhang[2]

1 Department of Chemistry and Chemical Engineering, Niigata University, Niigata, Japan

2 Graduate School of Science and Technology, Niigata University, Niigata, Japan

*Address all correspondence to: kano@eng.niigata-u.ac.jp

References

[1] Dutta S, Das AK. Analytical perspective on waste management for environmental remediation. Trends in Analytical Chemistry. 2010; **29**(7):636-644

[2] Sanchez A, Recillas S, Font X, et al. Ecotoxicity of, and remediation with, engineered inorganic nanoparticles in the environment. Trends in Analytical Chemistry. 2011;**30**(3):507-516

[3] Faraji M, Yamini Y, Rezaee M. Magnetic nanoparticles: Synthesis, stabilization, functionalization, characterization, and applications. Journal of the Iranian Chemical Society. 2010;7(1):1-37

[4] He M, Hu B, Zeng Y, Jiang Z. ICP-MS direct determination of trace amounts of rare earth impurities in various rare earth oxides with only one standard series. Journal of Alloys and Compounds. 2005;**390**:168-174

[5] Amirianshoja T, Junin R, Idris AK, Rahmani O. A comparative study of surfactant adsorption by clay minerals. Journal of Petroleum Science and Engineering. 2013;**101**:21-27

[6] Hatch CD, Wiese JS, Crane CC, et al. Water adsorption on clay minerals as a function of relative humidity: Application of BET and Freundlich adsorption models. Langmuir. 2012;**28**:1790-1803

[7] Zhao P, Liu X, Tian W, et al. Adsolubilization of 2, 4, 6-trichlorophenol from aqueous solution by surfactant intercalated ZnA layered double hydroxides. Chemical Engineering Journal. 2015;**279**:597-604

[8] Jiao FP, Chen XQ, Liu L, Hu ZD, Hu YH, Wang YH. Preparation and characterization of Mg-Al/Zn-Al layered double hydroxides intercalated with (+)-2, 3-di(p-toluyl)-tartaric acid. Journal of Molecular Structure. 2010;**964**(1):152-157

[9] Zhang H, Zou K, Guo SH, Duan X. Nanostructural drug-inorganic clay composites: Structure, thermal property and in vitro release of captopril-intercalated Mg-Al-layered double hydroxides. Journal of Solid State Chemistry. 2006;**179**(6):1792-1801

[10] Gyliene O, Aikaite J, Nivinskien O. Recovery of EDTA from complex solution using Cu(II) as precipitant and Cu(II) subsequent removal by electrolysis. Journal of Hazardous Materials. 2004;**116**(1):119-124

[11] Prieto C, Lozanoa JC, Rodríguez PB, Tome FV. Enhancing radium solubilization in soils by citrate, EDTA, and EDDS chelating amendments. Journal of Hazardous Materials. 2013;**250**:439-446

[12] Perez MR, Pavlovic I, Barriga C, et al. Uptake of Cu^{2+}, Cd^{2+} and Pb^{2+} on Zn-Al layered double hydroxide intercalated with EDTA. Applied Clay Science. 2006;**32**(3):245-251

[13] Gonzalez MA, Pavlovic I, Rojas R, Barriga C. Removal of Cu^{2+}, Pb^{2+} and Cd^{2+} by layered double hydroxide-humate hybrid. Sorbate and sorbent comparative studies. Chemical Engineering Journal. 2014;**254**:605-611

[14] Kaneyoshi M, Jones W. Formation of Mg-Al layered double hydroxides intercalated with nitrilotriacetate anions. Journal of Materials Chemistry. 1999;**9**:805

[15] Sui MH, Zhou YF, Sheng L, Duan BB. Adsorption of norfloxacin in aqueous solution by Mg-Al layered double hydroxides with variable metal composition and interlayer anions. Chemical Engineering Journal. 2012;**210**:451-460

[16] Liang XF, Hou WG, Xu YM, et al. Sorption of lead ion by layered double hydroxide intercalated with diethylenetriaminepentaacetic acid. Colloids and Surfaces A: Physicochemical and Engineering Aspects. 2010;**366**(1):50-57

[17] Escudero C, Gabaldon C, Marzal P, Villaescusa L. Effect of EDTA on divalent metal adsorption onto grape stalk and exhausted coffee wastes. Journal of Hazardous Materials. 2008;**152**:476-485

[18] Zhang S, Kano N, Imaizumi H. Adsorption of Cu(II), Pb(II) by Mg-Al layered double hydroxides (LDHs): Intercalated with the chelating agents EDTA and EDDS. Journal of Chemical Engineering of Japan. 2014;**47**(4):324-328

[19] Zhang S, Kano N, Imaizumi H. Adsorption of Cd(II) on Zn-Al LDHs (layered double hydroxides) intercalated with chelating agents EDTA. Journal of Chemistry and Chemical Engineering. 2016;**10**(2):60-67

[20] Zhang S, Kano N, Imaizumi H. Synthesis and characterization of LDHs (layered double hydroxides) intercalated with EDTA (ethylenediaminetetracetic acid) or EDDS (N, N'-1, 2-ethanediylbis-1-aspartic acid). Journal of Environmental Science and Engineering A. 2016;**5**:549-558

[21] Won SW, Choi SB, Yun YS. Performance and mechanism in binding of reactive orange 16 to various types of sludge. Biochemical Engineering Journal. 2006;**28**:208-214

[22] Wang B, Guo X, Peng B. Removal technology of boron dissolved in aqueous solutions—A review. Colloids and Surfaces A: Physicochemical and Engineering Aspects. 2014;**444**:338-344

[23] Ayawei N, Ekubo AT, Wankasi D, Dikio ED. Synthesis and application of layered double hydroxide for the removal of copper in wastewater. International Journal of Chemistry. 2015;7(1):122-132

[24] Ayawei N, Ekubo AT, Wankasi D, Dikio ED. Adsorption dynamics of copper adsorption by Zn/Al-CO_3. IJACSA. 2015;**3**(1):57-64

[25] Dolatyari L, Yaftian MR, Rostamnia S. Adsorption characteristics of Eu(III) and Th(IV) ions onto modified mesoporous silica SBA-15 materials. Journal of the Taiwan Institute of Chemical Engineers. 2016;**60**:174-184

[26] Gasser MS, Aly HF. Kinetic and adsorption mechanism of Cu(II) and Pb(II) on prepared nanoparticle layered double hydroxide intercalated with EDTA. Colloids and Surfaces A: Physicochemical and Engineering Aspects. 2009;**336**(1-3):167-173

[27] Rives V. Layered Double Hydroxides: Present and Future. New York: Nova Publishers; 2001

[28] Chitrakar R, Tezuka S, Sonoda A, et al. Synthesis and phosphate uptake behavior of Zr^{4+} incorporated MgAl-layered double hydroxides. Journal of Colloid and Interface Science. 2007;**313**:53-63

[29] Parida KM, Baliarsingh N, Sairam PB, Das J. Copper phthalocyanine immobilized Zn/Al LDH as photocatalyst under solar radiation for decolorization of methylene blue. Journal of Molecular Catalysis A: Chemical. 2006;**267**(1):202-208

[30] Park M, Choi CL, Seo YJ, et al. Reactions of Cu^{2+} and Pb^{2+} with Mg/Al layered double hydroxide. Applied Clay Science. 2007;**37**(1):143-148

[31] Cavani F, Trifirb F, Vaccari A. Hydrotalcite type anionic

clays: Preparation, properties and applications. Catalysis Today. 1991;**11**(2):173-301

[32] Rojas R. Copper, lead and cadmium removal by Ca Al layered double hydroxides. Applied Clay Science. 2014;**87**:254-259

[33] Osama S, Hideyuki T. Preparation and intercalation reactions of nano-structural materials, Zn-Al-Ti LDH. Materials Chemistry and Physics. 2008;**108**(2):449-455

[34] Zhang CL, Yu JY, Xu S, Xue LH, Cao ZL. Influence of UV aging on the rheological properties of bitumen modified with surface organic layered double hydroxides. Construction and Building Materials. 2016;**123**:574-580

[35] Liang XF, Zang YB, Xu YY, et al. Sorption of metal cations on layered double hydroxides. Colloids and Surfaces A: Physicochemical and Engineering Aspects. 2013;**433**:122-131

[36] Buxbaum G, Pfaff G. Cadmium Pigments. Industrial Inorganic Pigments. Hoboken, New Jersey, USA: Wiley-VCH; 2005. pp. 121-123

[37] Morrow H. Cadmium and Cadmium Alloys, Kirk-Othmer Encyclopedia of Chemical Technology. Hoboken, New Jersey, USA: John Wiley & Sons; 2010. pp. 1-36

[38] Kameda T, Saito S, Umetsu Y. Mg-Al layered double hydroxide intercalated with ethylene diaminetetraacetate anion: Synthesis and application to the uptake of heavy metal ions from an aqueous solution. Separation and Purification Technology. 2005;**47**(1):20-26

[39] Huang G, Wang D, Ma S, Chen J, Jiang L, Wang P. A new, low-cost adsorbent: Preparation, characterization, and adsorption behavior of Pb(II) and Cu(II). Journal of Colloid and Interface Science. 2015;**445**:294-302

[40] Anirudhan TS, Suchithra PS. Synthesis and characterization of tannin-immobilized hydrotalcite as a potential adsorbent of heavy metal ions in effluent treatments. Applied Clay Science. 2008;**42**:214-223

[41] Gonzalez MA, Pavlovic I, Barriga C. Cu(II), Pb(II) and Cd(II) sorption on different layered double hydroxides. A kinetic and thermodynamic study and competing factors. Chemical Engineering Journal. 2015;**269**:221-228

[42] Ma S, Chen Q, Li H, Wang P, Islam SM, Gu Q, et al. Highly selective and efficient heavy metal capture with polysulfide intercalated layered double hydroxides. Journal of Materials Chemistry A. 2014;**2**:10280-10289

[43] Zhou L, Ji L, Ma P-C, Shao Y, Zhang H, Gao W, et al. Development of carbon nanotubes/$CoFe_2O_4$ magnetic hybrid material for removal of tetrabromobisphenol A and Pb(II). Journal of Hazardous Materials. 2014;**265**:104-114

[44] Badruddoza AZM, Shawon ZBZ, Daniel TWJ, Hidajat K, Uddin MS. Fe_3O_4/cyclodextrin polymer nanocomposites for selective heavy metals removal from industrial wastewater. Carbohydrate Polymers. 2013;**91**:322-332

[45] Ma L, Wang Q, Islam SM, Liu Y, Ma S, Kanatzidis MG. Highly selective and efficient removal of heavy metals by layered double hydroxide intercalated with the MoS_4^{2-} ion. JACS. 2016;**138**:2858-2866

Section 3

Sorption Kinetics and Isotherm Models

Chapter 10

Modelling of Adsorption Kinetic Processes—Errors, Theory and Application

George William Kajjumba, Serkan Emik, Atakan Öngen, H. Kurtulus Özcan and Serdar Aydın

Abstract

Adsorption has become a competitive method in the field of wastewater and air treatment. Adsorption kinetics is one of the main factors that must be understood before the applicability of any adsorbent. In every adsorption process, linear or non-linear analysis of the kinetics is applied. The goodness of fit index (coefficient of correlation or sum of squares) is applied to access the best model. The usage of linear or non-linear from of the adsorption kinetics has an impact on the distribution of error function. Almost in every adsorption study, linear forms have been used to conclude the best kinetic model that influence the adsorption mechanism—which might be an error. Therefore, this review highlights the mistakes in the usage of linear and non-linear models. The applicability of the adsorption kinetics in wastewater treatment is also illuminated.

Keywords: goodness of fit, error function, Boyd model, diffusion model, dyes, ion adsorption, emerging contaminants

1. Introduction

Although water covers 70% of the world's surface, the availability of fresh water for animal consumption is becoming scarce. This is attributed to the improved standards that have resulted in rapid usage of pollutant infused substances such as pesticides, heavy metals, anions, pharmaceuticals, phenolic, personal care products, dyes, and hydrocarbons. Different methods have been developed to remove such substances from the wastewater, these include; biological, precipitation, membrane technology, electrochemical, and adsorption. Currently, the latter has received a considerable attention because it is cost and energy effective, easy to design and operate. Numerous adsorbents (carbon, clay, polymer, activates sludge, and zeolite) have been developed to remove solute form wastewater [1–3]. These adsorbents have large surface area and pore volume, thermal stability, with low acid/base reactivity. This makes them suitable to remove a wide range of organic and inorganic dissolved pollutants from wastewater and air.

The utmost parameter to consider while designing the adsorption system is adsorption kinetics; kinetics determine the rate at which the adsorption occurs. Kinetics are influenced by the surface complexity of the adsorbent, solute concentration and flow. Pseudo-First-order (PFO), Pseudo-Second-order (PSO), Elovich,

IntechOpen

and Intra-particle (IP) model are some of the kinetics that foretells the adsorbent-adsorbate interaction. The first two models have been widely applied in almost every sorption process. The suitability of any model depends on the error level—correlation coefficient (R^2) or Sum of Squared Errors (SSE). To study adsorption kinetics, the linear forms have been applied; a linear form of PSO has been favoured over PFO model for the last 2 decades.

The linearization of a function involves assumptions—any change in the assumption means a change in the error parameter. This leads to biasness in the data producing vast outcomes which violate the variance and normality assumption of the standard least squares. For example, the linearization of PSO has resulted in more than five different forms of PSO model. Using the non-linear form of the function means distributing the error on the entire curve. Generally, analysis of PFO and PSO is done by plotting $ln(q_t - q_e)$ and $t/_{q_t}$ against time, respectively. As the system approaches equilibrium, $t/_{q_t} \approx t/_{q_e}$ produces a perfect fitting for PSO. However, at equilibrium, $ln(q_e - q_t)$ becomes enormously large reducing the fitting index. In addition, R^2 has been compared using different scales of various kinetics—which is statically wrong. Therefore, this work highlights the gaps in the linearization of adsorption kinetics, and the suitability of each kinetic model towards the removal of dyes, organics, metals, and anions from solution.

2. Adsorption kinetics

Adsorption is the process by which solute molecules attach to the surface of an adsorbent. The adsorption process is done in batch or column setup. Adsorption kinetics is a curve (or line) that describes the rate of retention or release of a solute from an aqueous environment to solid-phase interface at a given adsorbents dose, temperature, flow rate and pH. During adsorption two main processes are involved; physical (physisorption) or chemical (chemisorption). Physical adsorption is as a result of weak forces of attraction (van der Waals), while chemisorption involves the formation of a strong bond between the solute and the adsorbent that involves the transfer of electrons.

2.1 Pseudo first order model (PFO)

Also known as Lagergren model, PFO describes the adsorption of solute onto adsorbent following the first order mechanism;

$$\frac{dq_t}{dt} = k_1(q_e - q_t) \tag{1}$$

where q_t is adsorbate adsorbed onto adsorbent at time t (mg/g), q_e is equilibrium adsorption capacity (mg/g), and k_1 is rate constant per min. The integral of Eq. (1) from $t = 0$ to $t = t$ and $q_t = 0$ and $q_t = q_t$ yields a linear expression of PFO, Eq. (2).

$$\ln(q_e - q_t) = lnq_e - k_1t \tag{2}$$

The value of k_1 is determined by plotting $\ln(q_e - q_t)$ vs. t. Albeit some studies have found k_1 to increase with initial solute concentration (C_0) or independent of C_0, the rate constant is always inversely proportional to the initial concentration of the solute. This is because a longer time is required for a large initial solute concentration. The controlling mechanism is affected by experimental conditions, thus the validity of the model. PFO is varied under Henry regime adsorption, and under high sorbent dosage [4]. To account for experimental differences during the sorption of dyes on activated carbon, PFO is modified as follows [5];

$$\frac{dq_t}{dt} = k_1 \frac{q_e}{q_t}(q_e - q_t) \tag{3}$$

and the linear form

$$\frac{q_t}{q_e} + \ln(q_e - q_t) = lnq_e - k_1 t \tag{4}$$

The plot of $\frac{q_t}{q_e} + \ln(q_e - q_t)$ against t gives a better fit of dye adsorption compared to Lagergren PFO. Different models for adsorption of various solutes have been developed, but the applicability and consistence of their linear and non-linear form need to be tested.

2.2 Pseudo second order (PSO) model

PSO model assumes that the rate of adsorption of solute is proportional to the available sites on the adsorbent. And the reaction rate is dependent on the amount of solute on the surface of the adsorbent—the driving force $(q_e - q_t)$ is proportional to the number of active sites available on the adsorbent [4, 6]. Eq. (5) shows the curvilinear form of PSO.

$$\frac{dq_t}{dt} = k_2 (q_e - q_t)^2 \tag{5}$$

k_2 is PSO rate constant. Applying the integral limits for t (0, t) and q_t $(0, q_t)$, the linearized form of PSO is;

$$q_t = \frac{t}{\frac{1}{k_2 q_e^2} + \frac{t}{q_e}} \tag{6}$$

Eq. (6) has been rearranged in different forms, Eq. (7)–(10).

$$\frac{t}{q_t} = \left[\frac{1}{k_2 q_e^2}\right] + \frac{t}{q_e} \tag{7}$$

$$\frac{1}{q_t} = \left[\frac{1}{k_2 q_e^2}\right]\frac{1}{t} + \frac{1}{q_e} \tag{8}$$

$$q_t = q_e - \left[\frac{1}{k_2 q_e}\right]\frac{q_t}{t} \tag{9}$$

$$\frac{q_t}{t} = k_2 q_e^2 - k_2 q_e q_t \tag{10}$$

The approach used to linearize the curvilinear function of PSO determines the distribution of the error function of the same kinetic model. Among the linearized equations of PSO, Eq. (7) yields better fitting results compared to other forms. Therefore, the PSO constants can be determined from a graph of t/q_t vs. t. Even though the PSO model may be affected by pH, dose amount, particle size, and temperature—the model assesses the impact of observable rate parameters.

PSO can be used to determine the initial solute uptake and adsorption capacity of an adsorbent. Within the last 2 decades, PSO fits the experiment better and it has been concluded that the adsorption mechanism is chemisorption in nature, involving the transfer of electrons between the adsorbate and adsorbent—this conclusion is wrong. Adsorption mechanism cannot be based on simple fitting of PSO model.

When the solute concentration is low, Eq. (7) explains the adsorption mechanism more than any other kinetic model; however, at high initial concentration, PFO model is favoured [7]. This is because at low C_0 the value of $\ln(q_e - q_t)$ increases exponentially increasing the error function—which is the reverse for high C_0. Although the applicable of linear forms has improved, they may be misleading in developing kinetic systems. For example, the R^2 of the linear PSO model during the adsorption of methylene green 5 onto activated was above 0.99; however, using non-linear form, the fit index was below 0.70 [6]. Both PSO and PFO do not explain the diffusion of solute into the adsorbent; therefore, before any conclusions are made about adsorption mechanism, diffusion models should be investigated.

2.3 Elovich model

To further understand the chemisorption nature of adsorption, Elovich model (developed by Zeldowitsch) is applied. This model helps to predict the mass and surface diffusion, activation and deactivation energy of a system. Although the model was initially applied in gaseous systems, its applicability in wastewater processes has been redeemed meaningful. The model assumes that the rate of adsorption of solute decreases exponentially as the amount of adsorbed solute increase.

$$\frac{dq_t}{dt} = \alpha \exp^{-\beta q_t} \tag{11}$$

As $q_t \approx 0$, $\frac{dq_t}{dt} \approx \alpha$ which is the initial adsorption rate (mg/g.min), and β is desorption constant. Integrating and applying the limits for t (0, t) and q_t $(0, q_t)$, the Elovich model can be linearized as;

$$q_t = \frac{1}{\beta}\ln\left[t + \frac{1}{\alpha\beta}\right] - \frac{1}{\beta}\ln(\alpha\beta) \tag{12}$$

As the system approaches equilibrium $t \gg \frac{1}{\alpha\beta}$, thus Eq. (12) becomes;

$$q_t = \frac{1}{\beta}\ln[\alpha\beta] + \frac{1}{\beta}\ln t \tag{13}$$

The graph of q_t *vs* t helps to determine the nature of adsorption on the heterogeneous surface of the adsorbent, whether chemisorption or not. A number of solutes have been reported to follow Elovich kinetics model [8, 9].

2.4 Intra-particle diffusion (IP) model

IP model has been widely applied to examine the rate limiting step during adsorption. The adsorption of solute in a solution involves mass transfer of adsorbate (film diffusion), surface diffusion, and pore diffusion. Film diffusion is an independent step, whereas surface and pore diffusion may occur simultaneously. IP is studied by examining Weber and Morris (1963) model, Eq. (14).

$$q_t = K_p\sqrt{t} + C \tag{14}$$

K_p is a rate constant mg/g.min$^{0.5}$, and C is boundary layer thickness. The values of C determines the boundary layer effect—higher values, the greater the effect. The plot of q_t *vs* $\sqrt{t}$, gives a linear function. If the line passes through the origin, IP diffusion controls the adsorption process. However, on many occasions, the plot does not pass through the origin and it gives multiple linear sections; these sections corresponds to different mechanisms that control the adsorption process. There are four main mechanisms that describe the transfer of solute from a solution to the adsorbent. The first is called mass transfer (bulk movement) of solute particles as soon as the adsorbent is dropped into the solution. This process is too fast, thus it is not considered during the design of kinetic systems. The second mechanism is called film diffusion; it involves the slow movement of solutes from the boundary layer to the adsorbent's surface. When the solute reach the surface of the adsorbent, they move to the pores of the adsorbent—third mechanism. The final mechanism involves rapid adsorptive attachment of the solute on the active sites of the pores; being a rapid process, it is not considered during engineering design of kinetics [6]. If the system is characterised by poor mixing, small solute size, and low concentration, film diffusion becomes the rate controlling step; otherwise, IP diffusion controls the process.

Misrepresentation of diffusion model: Couple of papers have assessed IP model using a straight of Eq. (14); however, in reality, pore diffusion is a slow process making Eq. (14) to follow a curvilinear trend. When the segment analysis is applied, the values of K_p and C differ enormously. The segments can be got by visual or application of regression on different points.

2.4.1 Boyd model

To understand if film diffusion is the rate controlling step, Boyd developed a single-resistance model that can be used to assess this effect. Boyd assumes that the boundary layer surrounding the adsorbent has a greater effect on the diffusion of solute [3]. To determine this effect, Eq. (15) is applied.

$$F = 1 - \frac{6}{\pi^2}\sum_{1}^{\infty}\left(\frac{1}{n^2}\right)\exp\left(-n^2 B_t\right) \tag{15}$$

F is the fraction of solute adsorbed at time t (q_t) to solute adsorbed at infinite time (q_∞) (t_∞ > $24h$ for better results). B_t is a mathematical function of F.

$$F = \frac{q_t}{q_\infty} \tag{16}$$

It is hard to estimate appropriate values of B_t with Eq. (15); B_t can be calculated using the integrated Fourier transform of Eqs. (17) and (18).

$$0 \le F \le 0.85:\ B_t = 2\pi - \frac{\pi^2 F}{3} - 2\pi\left(1 - \frac{\pi F}{3}\right)^{1/2} \tag{17}$$

$$0.86 \le F \le 1:\ B_t = -0.4977 - \ln(1 - F) \tag{18}$$

The graph of B_t *vs* t helps to predict the rate limiting step. If the graph approximates y = mx + 0.0 line, the rate limiting step is intra-particle diffusion, otherwise film diffusion model governs the process. Treatment of water was estimated to follow IP model [10, 11]. However, in many studies film diffusion is the limiting step during the initial stages of the process followed by IP diffusion when particles reach the surface of the adsorbent [1, 12, 13]. There is barely any non-linear form of IP model. A couple of published papers have mispresented Boyd model. The values of B_t are obtained using Eq. (18) over the entire time scale [11, 12], this is wrong.

3. Goodness of fit

To determine the kinetic model that best describes the interaction between the adsorbent and solute, the goodness of fit is used. The coefficient of correlation (R^2), sum-of-squared errors (SSE), average relative error, Spearman's correlation coefficient, non-linear chi-square test, hybrid fractional error function, Marquardt's percent standard deviation, and standard deviation of relative errors are some of the error functions that have been employed to study model fit. **Table 1** summaries the error function [4, 14–17].

3.1 Sum of squared errors

This is one of the most used error function in determining the model. The main challenge of using SSE is that at higher concentration the squares of error increase. This gives a good fit which is not the case always [15].

3.2 Coefficient of correlation

Almost given in every adsorption study, R^2 shows the degree of variability of dependent variable which is explained by all independent variables. It ranges from 0 to 1, with values close to zero showing a perfect fit.

3.3 Marquardt's percent standard deviation

Although not commonly used in adsorption kinetics analysis, this error function is a modification of geometric mean distribution. It is based on the number of degrees of freedom of a system [16].

Error function	Expression
Coefficient of correlation (R^2)	$\frac{\sum_{i=1}^{n}\left[q_{cal}-\overline{q_{exp}}\right]^2}{\sum_{i=1}^{n}\left[q_{cal}-\overline{q_{exp}}\right]^2+\sum_{i=1}^{n}\left[q_{cal}-q_{exp}\right]^2}$
Marquardt's percent standard deviation (MPSD)	$100\left[\sqrt{\frac{1}{n-p}\sum_{i=1}^{n}\left(\frac{q_{exp}-q_{cal}}{q_{exp}}\right)_i^2}\right]$
Sum-of-squared errors (SSE)	$\sum_{i=1}^{n}\left[q_{cal}-q_{exp}\right]_i^2$
Hybrid fractional error function	$\frac{100}{n-p}\sum_{i=1}^{p}\left[\frac{q_{exp}-q_{cal}}{q_{exp}}\right]_i$

q_{cal}: calculated amount of adsorbate adsorbed onto adsorbent, q_{exp}: experimental amount of adsorbate adsorbed, n: data points, p: number of parameters in each model.

Table 1.
Common error functions.

3.4 Hybrid fractional error function

Developed by Porter [17], the model was aimed to improve the applicability of SSE at a lower concentration. The error function is divided by the measured value.

3.5 Sum of normalised errors (SNE)

Different error functions yield different value of goodness fit—thus it may be difficult to select the best model fit. SNE provide a normalised value of the different error functions, making comparison very easy. SNE is done by dividing the error value of the different functions by the highest error for a given kinetic model.

3.6 Misuse of fitting index

The assessment of adsorption kinetics using error function has been misused in almost all adsorption papers. The problem arises when error function of linearized equations of non-linear functions are expended to determine the suitability of a model. In some linearized models, to reduce the error factor, log or square root transforms are applied if the error increases with the dependent factor. And if the error variance decrease with increasing dependent factor, then exponential or square alters are applied. However, the use of R^2 or SSE does not detect the biasness of the parameters.

The dependent variable in adsorption kinetic is not entirely linear over the given values of the independent variable. Eq. (7) shows the linearized form of PSO. The inverse of data weights $\left(1/q_t\right)$ and the presence of independent variable (t) in both dependent and independent sides causes false correlation. The inversing of variables on both sides of Eq. (8) distorts the error distribution over the entire data. In the third form of PSO (Eq. (9)), the presence of dependent parameter $\left(q_t\right)$ in both the independent and dependent section leads to spurious correlation. While in Eq. (10), the presence of independent variable violates the least squares assumption [18]. R^2 is a very sensitive parameter that can cause spurious conclusions. R^2 varies with the range of independent parameter—if the range is big, R^2 will be fit; and if the range is small, fit will be poor. Adding more data points decreases the degree of freedom of a system; this favours model fit. Therefore, making conclusions solely basing on R^2 can be misleading to the industry of adsorption mechanism.

4. Linear and non-linear fitting application

4.1 Dyes

Dyes are organic substances that cause a permanent or temporary change in colour of a material; they are resistant to detergents. Dyes are widely employed in leather, food, textile, paper, rubber, and plastic industries. When dyes are released in the hydrosphere, they can block sunlight penetration, thus affecting the marine life. In addition, they give unpleasant colour to water making it unsafe for human consumption. To reduce the impact of dyes on the ecosystem, adsorption method has been employed to remove dyes from wastewater. Different kinetic models have been employed to study the adsorption of dyes from solution, these include; PFO, PSO, Elvoich, and IP models. The suitability of any model depends on error functions. **Table 2** summarises the non-linear adsorption kinetics of different adsorbent.

Adsorbent	Solute	Error function	Kinetic	Ref.
Activated carbon	Acid red	R^2	PFO, PSO*	[19]
Activated carbon	Methylene blue	R^2	PFO*, PSO*	[20]
Activated carbon	Methylene blue	R^2, NSD	PFO*, PSO	[21]
Mn_2O_3 nanofibres	Methylene blue, malachite green, Basic red 46	R^2, X^2	PFO, PSO*, IP	[22]
Sugarcane bagasse	Auramine-O, Safranin-T	R^2, X^2	PFO, PSO*, IP, Elovich*	[23]
Polyhedral oligomeric	Crystal violet	R^2	PFO*, PSO*	[24]
MWCNTs hydrogel	Methylene blue	R^2, X^2, NSD	PFO*, PSO	[25]
$Mg_3Si_2O_5(OH)_4$	Rhodamine B	R^2	PFO*, PSO*	[26]
$Fe_3C/Fe_3O4/C$	Methylene blue	R^2	PFO, PSO*	[27]
Clinoptilolite	Pb	NSD, R^2, SSE	PFO*, PSO Elovich*	[28]
Clinoptilolite	Pb	R^2	PFO, PSO*	[29]
Wheat straw	Cu	R^2, X^2	PFO, PSO* Elovich	[30]
Fruit peels	Cu	R^2	PFO*, PSO Elovich	[31]
Chitosan	Cu	R^2, ARE	PFO, PSO* Elovich	[32]
Rape straw	Cu	R^2, X^2	PFO, PSO*, IP	[33]
Nano-TiO_2	As	NSD, R^2	PFO, PSO*	[34]
Nano-iron	As	NSD, R^2, SSE, ARE	PFO, PSO* Elovich	[35]
Graphene oxide	As	R^2, X^2	PFO*, PSO, IP, Elovich	[36]
Nanocellulose	Hg	R^2, X^2	PFO, PSO*	[37]
MCM-41	Hg	R^2, RMSE, ARE	PFO, PSO*, IP, Elovich	[38]
Almond shell	Hg	R^2, ARE	PFO, PSO*, IP,	[39]
Chitosan	P	R^2	PFO*, PSO*	[40]
Tourmaline	P	R^2	PFO, PSO*	[41]
Zn-Al	P	R^2, RMSE	PFO, PSO* IP	[42]
Iron	N	R^2, SSE	PFO*, PSO IP, Elovich	[43]
Chitosan-Fe	N	R^2, X^2, NSD	PFO, PSO* IP	[44]
Cellulose	N	R^2, ARE	PFO, PSO*, Avrami	[2]
Quintinite	F	R^2, X^2, SSE	PFO, PSO*, Elovich	[45]
Mg-Al-Fe	F	R^2, RMSE	PFO*, PSO, IP	[46]
Manganese carbonate	F	R^2, NSD	PFO, PSO*	[47]
Coconut	Carbofuran	R^2, ARE	PFO, PSO*	[48]
AC	Carbofuran	R^2, RMSE	PFO, PSO*	[49]

Adsorbent	Solute	Error function	Kinetic	Ref.
AC	Carbofuran	R^2	PFO, PSO*	[50]
AC	2,4-D	R^2, ARE, NSD	PFO, PSO*, IP	[51]
MIEX resin	2,4-D	R^2, RMSE	PFO, PSO*, IP	[52]
Sibunit	2,4-D	R^2	PFO*, PSO*,	[53]
AC	2,4-D	R^2, RMSE	PFO, PSO*	[54]
Peanut	Atrazine	R^2	PFO*, PSO*	[55]
MWCNT	Atrazine	R^2, ARE	PFO, PSO*, IP	[56]
Biochars	Atrazine	R^2	PFO, PSO*, IP, Elovich*	[57]
Trimethylsilylated	Gemfibrozil	R^2, SE	PFO, PSO*	[58]
Photocatalytic	Indomethacin	R^2, SE, SSE	PFO, PSO*	[59]
Human hair	Tetracycline	R^2, ARE	PFO*, PSO, IP	[60]

**Best model, NSD: normalised standard deviation, X^2 : chi-square, SE: standard error, SSE: sum of error squared, ARE: average relative error, RMSE: root-mean-square error, AC: activated carbon, 2,4-D: 2,4-dichlorophenoxyacetic acid.*

Table 2.
Non-linear adsorption kinetics.

While using a linear model of kinetics, based on R^2, it was evident the adsorption of Rhodamine B by $Mg_3Si_2O_5(OH)_4$ followed second order system; however, expending non-linear analysis, R^2 of PFO jumped from 0.67 to 0.99 [26]. Using non-linear approach to analyse the kinetics of dye adsorption, the goodness fit of PFO increases drastically to an average of 0.90. The fitting of $R^2 \geq 0.9$ is statistically good and can be used to make conclusions. Most error functions have been based on R^2—this has forced researchers to conclude that both PFO and PSO explain the adsorption of dye. However, further analysis using different error functions can help to solve the puzzle. The adsorption of Methylene blue onto activated carbon fitted PFO and PSO well ($R^2 > 0.9$); however, the analysis of NSE, PFO had the least value [21]—thus the adsorption of methylene blue is best described by PFO. Linearization of PFO worsens the adsorption parameter—to have a better fitting of PFO during dye adsorption, the non-linear model should be used while studying dye adsorption. Therefore as more dye adsorption studies are being carried, non-linear model and error functions should be explored to avoid misleading conclusions. Similar trends have been observed by different research groups working on dye adsorption [61–65].

4.2 Other organic materials

A number of organic material have been classified as emerging contaminates, these include; pesticide, pharmaceuticals, disinfection by-products, fertilisers, and biological toxins. All these organic materials affect our ecosystem in a diverse way. The cheap and effective method of removing these materials from the solution is adsorption; however, there has been mixed results about the adsorption kinetics of these pollutants. Carbofuran is a pesticide, ranked among the most dangerous carbamate insecticides. Albeit its bioaccumulation is less, its mammalian toxicity potential is high. When consumed for a prolonged period of time at elevated concentrations (more than 0.09 mg/L), it disrupts the endocrine, breathing, development and reproduction systems [50]. This necessitates its removal from wastewater. The adsorption kinetics of carbofuran have been studied in almost every adsorption study. The analysis of non-linear and linear kinetic functions using the error parameters—the adsorption of carbofuran generally follows PSO. The experimental data for adsorption of carbofuran on coconut were almost in perfect correlation with PSO model [48]. Therefore, PSO should be given a priority while designing adsorption system for carbofuran adsorption.

2,4-Dichlorophenoxyacetic acid (2,4-D) is herbicide that is globally used as a selective regulator of plants. 2,4-D is non-volatile but highly soluble in water. Thus exposure of groundwater to 2,4-D can lead to contamination. Long-time exposure of 100 μg/L of 2,4-D to humans can lead its accumulation in the seminal plasma and follicular mucus—increasing the risk of infertility. The removal of 2,4-D from solution using adsorption technique is mainly dominated by carbon related adsorbents [66]. Although the adsorption of 2,4-D using Sibunit raised a mixed observation as the initial concentration of changed [53], from **Table 2**, the assessment of error functions shows that adsorption of 2,4-D is mainly governed by PSO kinetics. Both non-linear and linear functions sorption kinetics favour PSO model. Another common herbicide expended in broadleaf weeds regulation is atrazine. Because of its high solubility in water, it has been detected in a number of groundwater wells, thus its ban in European countries; however, atrazine is extensively used in developing economies as pre and post-emergent herbicide to control weeds in crops. As 2,4-D, atrazine is an endocrine disruptor [56]. Several non-linear and linear kinetic studies have indicated that the adsorption of atrazine to be PSO in nature (**Table 2**).

Pharmaceuticals are organic compounds used to prevent, treat, and restore organ function. After restoring the damaged organ, they are flashed out of the body into the environment. In addition, pharmaceutical companies release numerous amount of waste into the water streams. The widely employed wastewater treatment technologies cannot remove pharmaceuticals wastes—thus the escalated presence of pharmaceutical products in the environment. Estrone, 17β-estradiol, 17α-ethinylestradiol, estriol, and acetaminophen are among the most alkaloids, steroid hormones and primary estrogens present in wastewater. As pesticides and herbicides, pharmaceuticals danger the nervous system. Most of the pharmaceuticals that are harmful to human nervous system tend to be lipophilic in nature, binding to solids. Nanofiltration, advanced oxidation, and adsorption are some of the main methods that are used to remove pharmaceuticals from wastewater. To effectively remove pharmaceuticals from solution, mesopore adsorbents with considerable surface area must be used. The adsorption kinetics of anionic, neutral, and cationic pharmaceutical onto mesopore trimethylsilylated all followed PSO model [58].

4.3 Metals and anions

In allowable range, metals are vital elements that support human life—without them, life could be impossible. Among all the major contaminants, metals top the list because of their wide applicability in electricity, construction, and medicine. Accumulation of heavy metals in human bodies is dangerous. For example, thallium and manganese damages nervous system, while cobalt and nickel are carcinogenic. Different adsorbents have been developed to remove these metals from wastewater. These include; industrial waste, peat, wood, brown rice, straw, peanut shell, hazelnut shell, soybean filament, cotton seed pulp, sugar beet pulp, leaves. The adsorption of metals is affected by pH of the solution, contact time, the initial concentration of metal and ambient temperature are affected. The adsorption mechanism is evaluated by various kinetic models such as PFO, PSO, Elovich,

Weber and Morris intra-particle diffusion model, Bangham model and Boyd kinetic model.

During metal adsorption, the best kinetic model is evaluated by assessing the error function after plotting linear or and linear model. In addition, qe must match reasonably well with experimental values, qe_{cal} at all initial concentrations of adsorbate with maximum R^2 and minimum χ^2 values. The adsorption of lead by pre-treated clinoptilolite using the non-linear kinetics was best described by PFO [28]. Both raw and treated clinoptilolite PFO had R^2 above 0.98 in addition to having the least values of SSE—the experimental values agreed well with the calculated values. The rate constant increased with the increase in the initial solute concentration. Using almost similar adsorbent, bentonite to adsorbed Pb from solution was best described by PSO than any other kinetic model [29]. This observation was based on the linear format of the kinetics—therefore, to make conclusive meaning, both non-linear and linear model should be investigated.

Copper is another mineral that has received considerable attention over the past decades. Copper nanoparticles have been employed in lubricates to reduce the friction, tear and mend torn surfaces. Because of its good charge-discharge property, copper nanoparticles have been used in lithium batteries to improve coulombs efficiency. Due to numerous application of copper, it is mostly likely to enter into human bodies via the food and water chain. Excessive exposer of animals to copper destruct lipid profile, malfunction of the renal, and hepatocirrhosis [33]; this calls its removal from solution before it is consumed by humans. The removal of copper by adsorption process has been studied by different research groups. Although there are some studies that have favoured the adsorption of copper to be PFO nature [31], the assessment non-linear and linear adsorption kinetics of copper favours PSO—regardless of the adsorbent (**Table 2**). Comparing the error functions that determine the suitability of the kinetic model, PSO error parameters possess minimal variations. For example, the adsorption of copper onto chitosan followed PSO mechanism [32].

Ranked in top five among the most dangerous metal, the occurrence of Arsenic in wastewater is on raise due to increased usage of paintings, dyes, mining and smelting activities. The consumption of arsenic contaminated materials (above 10 parts per billion) causes muscle cramping, blood and hair loss—thus it is necessary to remove arsenic from water before it is consumed. The adsorption kinetics of arsenic vary from one experiment to another. While comparing the non-linear and linear kinetics of arsenic adsorption by nano-TiO_2, both non-linear form of PFO and PSO yielded good fit (>0.90) [34]. The linear model of PFO yielded a very poor fit—thus basing on the linear model only can lead to poor judgement. Although there is no clear cut point on the best kinetic model, the recent adsorption studies of arsenic favour PSO (**Table 2**). Mercury is another heavy metal that has been investigated extensively due to its toxicity to animals. Although Hg is widely used in teeth amalgam, consumption of water levels with more 0.01 mg/L of Hg causes neuronal disorders and damages cardiovascular system [67]. This has called the development of cheap and efficient adsorbents to remove Hg from solution. While modelling the linear adsorption of Hg using MCM-41, PFO yielded a poor coefficient; however, when non-linear models were applied, PFO error functions improved [38]. **Table 2** summaries the adsorption kinetics of Hg onto different adsorbents—the non-linear PSO model has been preferred over the other models.

Metals in wastewater are always accompanied by anions like phosphate, sulphate, carbonate, and chloride. Among these, phosphate is the most crucial element. Phosphorus is a vital element in our ecosystem, without it life would be impossible. Phosphorus support bone and tooth growth, nerves and muscles. Phosphorus is vital during the formation and maintenance of DNA and cell membranes. However, excessive presence of phosphorus in the ecosystem promotes bacteria growth. This affects marine life which in turn disorganises the ecosystem. Also, white phosphorus has been reported to cause stomach cramps, nausea, and drowsiness [68]. Various approaches are used in phosphate removal from wastewater including ion exchange, chemical precipitation, biological, and adsorption. The later has received considerable attention due to its efficiency and cost benefit. The adsorption kinetics of vanadium have been extensively studied in almost every study. There are mixed results about the adsorption kinetics of phosphorus. The adsorption of phosphorus onto tourmaline was best described by linear PSO; however, analysis of the non-linear models produced better fitting for both models ($R^2 > 0.95$) [41]. Thus, before any conclusion is made about the kinetic model, non-linear kinetics should be investigated. There is no clear model that governs the adsorption of phosphorus (**Table 2**).

Nitrates are another ions that affect ecosystem extensively. The mechanisation of agriculture has promoted the use of nitrogen infused fertilisers at large scale. Excessive nitrogen promotes eutrophication—a condition that promotes algal growth. Algal growth cuts off oxygen supply in the aquatic system [69]. Therefore, before nitrate infused wastewater is released into the atmosphere, it must be treated. In addition, elevated amount of nitrate cause blue-baby syndrome. Adsorption has been employed to remove nitrates from wastewater. The adsorption of nitrate onto iron particles was best described by PFO [43]. However, during the adsorption of nitrate onto chitosan-Fe, PSO was favoured model [44]. There is more need to investigate the non-linear kinetics of nitrate adsorption. Fluorine is an essential element in our daily life—it is used in toothpaste to prevent teeth from decay. However, long term consumption of water with over 1.5 mg/L causes fluorosis, a condition that affects teeth, bones ossification, and neurological damage under extreme conditions [70]. To remove fluorine from water, precipitation and adsorption have been employed extensively. The adsorption kinetics of fluoride are complex, they depend on solute-adsorbent interaction. For example, the adsorption of fluoride onto manganese carbonate was second order in nature [47]; however, the analysis of a non-linear model of fluoride adsorption by Mg-Al-Fe, PFO controlled the reaction [46].

5. Conclusion

Since the late C20th, majority of the adsorption studies have favoured PSO than PFO. This is attributed to the fact that, most of the plots of PSO include values as the system approaches equilibrium—the values of $^{t}/_{q_t} \approx {}^{t}/_{q_e}$. The incorporation of values close to equilibrium produces a fitting index close to one. For PFO, as the system approaches equilibrium, the $(q_t - q_e)$ slant to zero, thus $ln(q_e - q_t)$ becomes abnormally large at equilibrium reducing the accuracy. In all adsorption studies sampled, R^2 has is used to test goodness fit. However, to have a better comparison of R^2, the scale must be the same. Therefore, to understand the adsorption mechanism of any solute, non-linear models should be applied, and to assess the best fit model, non-linear least squares must be applied. Many linear forms of PFO, PSO and Elovich have been developed, but most of them are based on erroneous assumptions. Consequently, while studying the adsorption kinetics, the above given equations should be used. Using PSO model to conclude that the adsorption kinetics is chemisorption is misleading. The models should also be checked with diffusion models to best describe the adsorption mechanism.

Acknowledgements

This study was funded by Istanbul University with project number BYP-2017-22921. Thanks to Joseph Wasswa, Laura Milillo and Michael Kayemba of UZURI Advisory for the suggestions and advice.

Author details

George William Kajjumba[1], Serkan Emik[2], Atakan Öngen[1], H. Kurtulus Özcan[1] and Serdar Aydın[1]*

1 Department of Environmental Engineering, Istanbul University, Istanbul, Turkey

2 Department of Chemical Engineering, Istanbul University, Istanbul, Turkey

*Address all correspondence to: saydin@istanbul.edu.tr

References

[1] Emik S. Preparation and characterization of an IPN type chelating resin containing amino and carboxyl groups for removal of Cu(II) from aqueous solutions. Reactive and Functional Polymers. 2014;**75**(1):63-74

[2] Anirudhan TS, Rauf TA. Adsorption performance of amine functionalized cellulose grafted epichlorohydrin for the removal of nitrate from aqueous solutions. Journal of Industrial and Engineering Chemistry. 2013;**19**(5): 1659-1667

[3] Boyd GE, Schubert J, Adamson AW. The exchange adsorption of ions from aqueous solutions by organic zeolites. Ion-exchange equilibria. Journal of the American Chemical Society. 1947; **69**(11):2818-2829

[4] Tan KL, Hameed BH. Insight into the adsorption kinetics models for the removal of contaminants from aqueous solutions. Journal of the Taiwan Institute of Chemical Engineers. 2017; **74**:25-48

[5] Yang X, Al-Duri B. Kinetic modeling of liquid-phase adsorption of reactive dyes on activated carbon. Journal of Colloid and Interface Science. 2005; **287**(1):25-34

[6] Tran HN, You SJ, Hosseini-Bandegharaei A, Chao HP. Mistakes and inconsistencies regarding adsorption of contaminants from aqueous solutions: A critical review. Water Research. 2017; **120**:88-116

[7] Azizian S. Kinetic models of sorption: A theoretical analysis. Journal of Colloid and Interface Science. 2004;**276**(1): 47-52

[8] Dotto GL, Pinto LAA. Adsorption of food dyes onto chitosan: Optimization process and kinetic. Carbohydrate Polymers. 2011;**84**(1):231-238

[9] Cheung CW, Porter JF, McKay G. Elovich equation and modified second-order equation for sorption of cadmium ions onto bone char. Journal of Chemical Technology and Biotechnology. 2000; **75**(11):963-970

[10] Singh SK, Townsend TG, Mazyck D, Boyer TH. Equilibrium and intra-particle diffusion of stabilized landfill leachate onto micro- and meso-porous activated carbon. Water Research. 2012; **46**(2):491-499

[11] Elkady MF, Ibrahim AM, El-Latif MMA. Assessment of the adsorption kinetics, equilibrium and thermodynamic for the potential removal of reactive red dye using eggshell biocomposite beads. Desalination. 2011;**278**(1–3):412-423

[12] Tang H, Zhou W, Zhang L. Adsorption isotherms and kinetics studies of malachite green on chitin hydrogels. Journal of Hazardous Materials. 2012;**209–210**:218-225

[13] Kajjumba GW, Aydın S, Güneysu S. Adsorption isotherms and kinetics of vanadium by shale and coal waste. Adsorption Science & Technology. 2018;**36**(3–4):936-952. Available from: http://journals.sagepub.com/doi/10.1177/0263617417733586

[14] Subramanyam B, Das A. Linearised and non-linearised isotherm models optimization analysis by error functions and statistical means. Journal of Environmental Health Science & Engineering. 2014;**12**(1):1-6

[15] Foo KY, Hameed BH. Insights into the modeling of adsorption isotherm systems. Chemical Engineering Journal. 2010;**156**:2-10

[16] Demirbas E, Kobya M, Konukman AES. Error analysis of equilibrium studies for the almond shell activated carbon adsorption of Cr(VI) from aqueous solutions. Journal of Hazardous Materials. 2008;**154**(1–3):787-794

[17] Porter JF, McKay G, Choy KH. The prediction of sorption from a binary mixture of acidic dyes using single- and mixed-isotherm variants of the ideal adsorbed solute theory. Chemical Engineering Science. 1999;**54**(24): 5863-5885

[18] El-Khaiary MI, Malash GF. Common data analysis errors in batch adsorption studies. Hydrometallurgy. 2011;**105**(3–4):314-320

[19] Salman Naeem M, Javed S, Baheti V, Wiener J, Javed MU, Ul Hassan SZ, et al. Adsorption kinetics of acid red on activated carbon web prepared from acrylic fibrous waste. Fibers and Polymers. 2018;**19**(1):71-81. Available from: http://link.springer.com/10.1007/s12221-018-7189-5

[20] Aichour A, Zaghouane-Boudiaf H, Iborra CV, Polo MS. Bioadsorbent beads prepared from activated biomass/alginate for enhanced removal of cationic dye from water medium: Kinetics, equilibrium and thermodynamic studies. Journal of Molecular Liquids. 2018;**256**:533-540

[21] Baysal M, Bilge K, Yılmaz B, Papila M, Yürüm Y. Preparation of high surface area activated carbon from waste-biomass of sunflower piths: Kinetics and equilibrium studies on the dye removal. Journal of Environmental Chemical Engineering. 2018;**6**(2): 1702-1713

[22] Berenjian A, Maleknia L, Chizari Fard G, Almasian A. Mesoporous carboxylated Mn_2O_3 nanofibers: Synthesis, characterization and dye removal property. Journal of the Taiwan Institute of Chemical Engineers. 2018: 1-16. DOI: 10.1016/j.jtice.2018.01.050

[23] Fideles RA, Ferreira GMD, Teodoro FS, Adarme OFH, da Silva LHM, Gil LF, et al. Trimellitated sugarcane bagasse: A versatile adsorbent for removal of cationic dyes from aqueous solution. Part I: Batch adsorption in a monocomponent system. Journal of Colloid and Interface Science. 2018;**515**:172-188

[24] Eftekhari-Sis B, Akbari A, Motlagh PY, Bahrami Z, Arsalani N. Dye adsorption on cubic polyhedral oligomeric silsesquioxane-based poly (acrylamide-co-itaconic acid) hybrid nanocomposites: Kinetic, thermodynamic and isotherms studies. Journal of Inorganic and Organometallic Polymers and Materials. 2018;**28**(5): 1728-1738

[25] Makhado E, Pandey S, Nomngongo PN, Ramontja J. Preparation and characterization of xanthan gum-cl-poly (acrylic acid)/o-MWCNTs hydrogel nanocomposite as highly effective re-usable adsorbent for removal of methylene blue from aqueous solutions. Journal of Colloid and Interface Science. 2018;**513**:700-714

[26] Sun P, Xu L, Li J, Zhai P, Zhang H, Zhang Z, et al. Hydrothermal synthesis of mesoporous $Mg_3Si_2O_5(OH)$ 4microspheres as high-performance adsorbents for dye removal. Chemical Engineering Journal. 2018;**334**:377-388

[27] Yao L, Yang J, Zhang P, Deng L. In situ surface decoration of Fe3C/Fe3O4/C nanosheets: Towards bi-functional activated carbons with supercapacitance and efficient dye adsorption. Bioresource Technology. 2018;**256**:208-215

[28] Günay A, Arslankaya E, Tosun I. Lead removal from aqueous solution by natural and pretreated clinoptilolite:

Adsorption equilibrium and kinetics. Journal of Hazardous Materials. 2007; **146**(1–2):362-371

[29] Ahmad R, Mirza A. Synthesis of Guar gum/bentonite a novel bionanocomposite: Isotherms, kinetics and thermodynamic studies for the removal of Pb (II) and crystal violet dye. Journal of Molecular Liquids. 2018;**249**: 805-814

[30] Han R, Zhang L, Song C, Zhang M, Zhu H, Zhang L. Characterization of modified wheat straw, kinetic and equilibrium study about copper ion and methylene blue adsorption in batch mode. Carbohydrate Polymers. 2010; **79**(4):1140-1149

[31] Romero-Cano LA, García-Rosero H, Gonzalez-Gutierrez LV, Baldenegro-Pérez LA, Carrasco-Marín F. Functionalized adsorbents prepared from fruit peels: Equilibrium, kinetic and thermodynamic studies for copper adsorption in aqueous solution. Journal of Cleaner Production. 2017;**162**:195-204

[32] Frantz TS, Silveira N, Quadro MS, Andreazza R, Barcelos AA, Cadaval TRS, et al. Cu(II) adsorption from copper mine water by chitosan films and the matrix effects. Environmental Science and Pollution Research. 2017; **24**(6):5908-5917

[33] Liu X, Chen ZQ, Han B, Su CL, Han Q, Chen WZ. Biosorption of copper ions from aqueous solution using rape straw powders: Optimization, equilibrium and kinetic studies. Ecotoxicology and Environmental Safety. 2018;**150**:251-259

[34] Yazdani MR, Bhatnagar A, Vahala R. Synthesis, characterization and exploitation of nano-TiO_2/feldspar-embedded chitosan beads towards UV-assisted adsorptive abatement of aqueous arsenic (As). Chemical Engineering Journal. 2017;**316**:370-382

[35] Asmel NK, Yusoff ARM, Sivarama Krishna L, Majid ZA, Salmiati S. High concentration arsenic removal from aqueous solution using nano-iron ion enrich material (NIIEM) super adsorbent. Chemical Engineering Journal. 2017;**317**:343-355

[36] Maji S, Ghosh A, Gupta K, Ghosh A, Ghorai U, Santra A, et al. Efficiency evaluation of arsenic(III) adsorption of novel graphene oxide@iron-aluminium oxide composite for the contaminated water purification. Separation and Purification Technology. 2018;**197** (September 2017):388-400. DOI: 10.1016/j.seppur.2018.01.021

[37] Anirudhan TS, Shainy F. Effective removal of mercury(II) ions from chlor-alkali industrial wastewater using 2-mercaptobenzamide modified itaconic acid-grafted-magnetite nanocellulose composite. Journal of Colloid and Interface Science. 2015;**456**:22-31

[38] Raji F, Pakizeh M. Kinetic and thermodynamic studies of Hg(II) adsorption onto MCM-41 modified by $ZnCl_2$. Applied Surface Science. 2014: 568-575

[39] Taha AA, Moustafa AHE, Abdel-Rahman HH, Abd El-Hameed MMA. Comparative biosorption study of Hg (II) using raw and chemically activated almond shell. Adsorption Science and Technology. 2018;**36**(1–2):521-548

[40] Liu X, Zhang L. Removal of phosphate anions using the modified chitosan beads: Adsorption kinetic, isotherm and mechanism studies. Powder Technology. 2015;**277**:112-119

[41] Li G, Chen D, Zhao W, Zhang X. Efficient adsorption behavior of phosphate on La-modified tourmaline. Journal of Environmental Chemical Engineering. 2015;**3**(1):515-522

[42] Hatami H, Fotovat A, Halajnia A. Comparison of adsorption and

desorption of phosphate on synthesized Zn-Al LDH by two methods in a simulated soil solution. Applied Clay Science. 2018;**152**:333-341

[43] Kim DG, Hwang YH, Shin HS, Ko SO. Kinetics of nitrate adsorption and reduction by nano-scale zero valent iron (NZVI): Effect of ionic strength and initial pH. KSCE Journal of Civil Engineering. 2016;**20**(1):175-187

[44] Hu Q, Chen N, Feng C, Hu W. Nitrate adsorption from aqueous solution using granular chitosan-Fe^{3+} complex. Applied Surface Science. 2015; **347**:1-9. Available from: http://linkinghub.elsevier.com/retrieve/pii/S0169433215008892

[45] Kim JH, Park JA, Kang JK, Son JW, Yi IG, Kim SB. Characterization of quintinite particles in fluoride removal from aqueous solutions. Environmental Engineering Research. 2014;**19**(3):247-253

[46] Ma W, Zhao N, Yang G, Tian L, Wang R. Removal of fluoride ions from aqueous solution by the calcination product of Mg-Al-Fe hydrotalcite-like compound. Desalination. 2011;**268**(1–3):20-26

[47] Zhang YX, Jia Y. Fluoride adsorption on manganese carbonate: Ion-exchange based on the surface carbonate-like groups and hydroxyl groups. Journal of Colloid and Interface Science. 2018;**510**:407-417

[48] Njoku VO, Islam MA, Asif M, Hameed BH. Preparation of mesoporous activated carbon from coconut frond for the adsorption of carbofuran insecticide. Journal of Analytical and Applied Pyrolysis. 2014;**110**(1):172-180

[49] Salman JM, Njoku VO, Hameed BH. Bentazon and carbofuran adsorption onto date seed activated carbon: Kinetics and equilibrium. Chemical Engineering Journal. 2011;**173**(2):361-368. DOI: 10.1016/j.cej.2011.07.066

[50] Chang KL, Chen CC, Lin JH, Hsien JF, Wang Y, Zhao F, et al. Rice straw-derived activated carbons for the removal of carbofuran from an aqueous solution. Xinxing Tan Cailiao/New Carbon Materials. 2014;**29**(1):47-54

[51] Njoku VO, Islam MA, Asif M, Hameed BH. Adsorption of 2,4-dichlorophenoxyacetic acid by mesoporous activated carbon prepared from H_3PO_4-activated langsat empty fruit bunch. Journal of Environmental Management. 2015;**154**:138-144

[52] Lu X, Shao Y, Gao N, Ding L. Equilibrium, thermodynamic, and kinetic studies of the adsorption of 2,4-dichlorophenoxyacetic acid from aqueous solution by MIEX resin. Journal of Chemical & Engineering Data. 2015; **60**(5):1259-1269. Available from: http://pubs.acs.org/doi/abs/10.1021/je500902p

[53] Vedenyapina MD, Sharifullina LR, Kulaishin SA, Strel'tsova ED, Vedenyapin AA, Lapidus AL. Adsorption of 2,4-dichlorophenoxyacetic acid and phenoxyacetic acid on sibunit. Solid Fuel Chemistry. 2018; **52**(1):53-57

[54] Tang L, Zhang S, Zeng GM, Zhang Y, De Yang G, Chen J, et al. Rapid adsorption of 2,4-dichlorophenoxyacetic acid by iron oxide nanoparticles-doped carboxylic ordered mesoporous carbon. Journal of Colloid and Interface Science. 2015;**445**:1-8

[55] Saha A, Bhaduri D, Pipariya A, Kumar GR. Linear and nonlinear sorption modelling for adsorption of atrazine onto activated peanut husk. Environmental Progress & Sustainable Energy. 2017;**36**(2):348-358

[56] Chen GC, Shan XQ, Zhou YQ, Shen XE, Huang HL, Khan SU. Adsorption kinetics, isotherms and thermodynamics of atrazine on surface oxidized multiwalled carbon nanotubes. Journal of Hazardous Materials. 2009; **169**(1–3):912-918

[57] Mandal A, Singh N, Purakayastha TJ. Characterization of pesticide sorption behaviour of slow pyrolysis biochars as low cost adsorbent for atrazine and imidacloprid removal. Science of the Total Environment. 2017; **577**:376-385

[58] Bui TX, Pham VH, Le ST, Choi H. Adsorption of pharmaceuticals onto trimethylsilylated mesoporous SBA-15. Journal of Hazardous Materials. 2013; **254–255**(1):345-353

[59] Basha S, Keane D, Morrissey A, Nolan K, Oelgemöller M, Tobin J. Studies on the adsorption and kinetics of photodegradation of pharmaceutical compound, indomethacin using novel photocatalytic adsorbents (IPCAs). Industrial and Engineering Chemistry Research. 2010;**49**(22):11302-11309

[60] Ahmed MJ, Islam MA, Asif M, Hameed BH. Human hair-derived high surface area porous carbon material for the adsorption isotherm and kinetics of tetracycline antibiotics. Bioresource Technology. 2017;**243**:778-784

[61] Mall ID, Srivastava VC, Agarwal NK. Removal of orange-G and methyl violet dyes by adsorption onto bagasse fly ash - kinetic study and equilibrium isotherm analyses. Dyes and Pigments. 2006;**69**(3):210-223

[62] Lin J, Wang L. Comparison between linear and non-linear forms of pseudo-first-order and pseudo-second-order adsorption kinetic models for the removal of methylene blue by activated carbon. Frontiers of Environmental Science & Engineering in China. 2009; **3**(3):320-324

[63] Crini G, Peindy HN, Gimbert F, Robert C. Removal of C.I. basic green 4 (malachite green) from aqueous solutions by adsorption using cyclodextrin-based adsorbent: Kinetic and equilibrium studies. Separation and Purification Technology. 2007;**53**(1): 97-110

[64] Crini G. Kinetic and equilibrium studies on the removal of cationic dyes from aqueous solution by adsorption onto a cyclodextrin polymer. Dyes and Pigments. 2008;**77**(2):415-426

[65] Chowdhury S, Saha P. Adsorption kinetic modeling of safranin onto rice husk biomatrix using pseudo-first- and pseudo-second-order kinetic models: Comparison of linear and non-linear methods. Clean: Soil, Air, Water. 2011; **39**(3):274-282

[66] Bongiovann B, Konjuh C, Pochettino A, Ferri A. Oxidative stress as a possible mechanism of toxicity of the herbicide 2,4-dichlorophenoxyacetic acid (2,4-D). Herbicides. 2012:315-334

[67] Kobal AB, Kobal GD. Elemental mercury exposure and sleep disorder. Sleep Disorders. 2012:47-64

[68] Budnikova YG, Krasnov SA. Electrochemical transformation of white phosphorus as a way to compounds with phosphorus-hydrogen and phosphorus-carbon bonds. Developments in Electrochemistry. 2012:101-124

[69] Simonič M, Goršek A, Petrovič A. Nitrate removal from groundwater with membrane bioreactor. Nitrification and Denitrification. 2017:93-109. Available from: http://www.intechopen.com/books/nitrification-and-denitrification/nitrate-removal-from-groundwater-with-membrane-bioreactor

[70] Thole B. Ground water contamination with fluoride and potential fluoride removal technologies

for east and southern Africa. Perspectives in Water Pollution. 2013: 65-95. Available from: http://www.intechopen.com/books/perspectives-in-water-pollution/ground-water-contamination-with-fluoride-and-potential-fluoride-removal-technologies-for-east-and-so